THE GRAIN BRAIN WHOLE LIFE PLAN

ALSO BY DR DAVID PERLMUTTER

Brain Maker: The Power of Gut Microbes to Heal and Protect Your Brain — for Life

The Grain Brain Cookbook: More Than 150 Life-Changing Gluten-Free Recipes to Transform Your Health

Grain Brain: The Surprising Truth about Wheat, Carbs, and Sugar — Your Brain's Silent Killers

Power Up Your Brain

Raise a Smarter Child by Kindergarten

The Better Brain Book

THE GRAIN BRAIN WHOLE LIFE PLAN

Boost Brain Performance, Lose Weight, and Achieve Optimal Health

BY DR DAVID PERLMUTTER

WITH KRISTIN LOBERG

This book is intended to supplement, not replace, the advice of a trained health professional. If you know or suspect that you have a health problem, you should consult a health professional. The author and publisher specifically disclaim any liability, loss, or risk, personal or otherwise, that is incurred as a consequence, directly or indirectly, of the use and application of any of the contents of this book.

First published in Great Britain in paperback in 2017

First published in Great Britain in 2016 by Yellow Kite
An imprint of Hodder & Stoughton
An Hachette UK company

First published in the USA in 2016 by Little, Brown and Company

A CIP catalogue record for this title is available from the British Library

Paperback ISBN 978 1 473 64779 4
eBook ISBN 978 1 473 64778 7

Printed and bound by Clays Ltd, St Ives plc

Hodder & Stoughton policy is to use papers that are natural, renewable and recyclable products and made from wood grown in sustainable forests. The logging and manufacturing processes are expected to conform to the environmental regulations of the country of origin.

Yellow Kite
Hodder & Stoughton Ltd
Carmelite House
50 Victoria Embankment
London EC4Y 0DZ

www.yellowkitebooks.co.uk

This book is dedicated to my wife, Leize.
To be blessed by your love is the brightest light of my life.

Contents

THE GRAIN BRAIN WHOLE LIFE PLAN

INTRODUCTION

You've Come to This Book for a Reason

You can choose your health destiny. Whether it's effortless weight loss. Freedom from neurological disorders and other chronic conditions. Boundless energy. A radiant appearance. Sound sleep. A happy belly. A robust immune system. Relief from depression and anxiety. A sharp, fast-thinking brain. A great sense of self-confidence and well-being. A super-high quality of life....

These are all terrific goals, and my bet is you're hoping to achieve them yourself soon enough. People who've followed my protocols in the past have indeed experienced these results. Seriously. But to be sure: Such monumental accomplishments don't come without hard work and sacrifices. You can't necessarily turn away from a standard Western diet—bread, fizzy drinks, OJ, sugar, cereal, muffins, bagels, processed foods—and embrace a totally gluten-free, low-carb lifestyle overnight. It takes commitment. It takes effort. But it's doable with this book in your personal library.

More than a million people around the world have improved their health—physical, mental, and cognitive—thanks to *Grain Brain*, a book that became an instant bestseller. The book was followed by *Brain Maker*, another instant bestseller that added to the conversation by highlighting the importance of the human microbiome—the trillions of microbes that inhabit the gut—to our health. Now the time has come to bring these two forces together in a highly practical, step-by-step holistic lifestyle program.

Welcome to *The Grain Brain Whole Life Plan.*

The main purpose of this book is to help you put my ideas into practice in the real world and to show you that living your best life is about much more than what you put in your mouth. It expands upon the core advice in my previous works and introduces exciting new information about the advantages of eating more fat and fiber, consuming fewer carbs and protein, evicting gluten forever, and catering to your intestinal flora. Included in the book are a bounty of delicious original recipes, tips for addressing unique challenges, a 14-day easy-to-follow meal plan, and advice about habits beyond the diet. From sleep hygiene to stress management, exercise, supplements, and more, *The Grain Brain Whole Life Plan* details how to live happily and healthily ever after.

Grain Brain and *Brain Maker* share the foundation of my general nutritional recommendations, complete with all the scientific evidence supporting them. I highly recommend you read these books if you haven't already before tackling this program. They tell you the WHY part of the story in rich detail. *The Grain Brain Whole Life Plan* offers the HOW. If you've read my previous works, you'll hear some echoes in these pages, but that's intentional. The reminders will reinforce your motivation to change, or to keep up the good work.

My ideas may have seemed out there when I began writing *Grain Brain* in 2012, but since then, not only have they been validated over and over again in the scientific literature, but also more extensive science has emerged, which I'll address in this work. Even the U.S. government has modified its dietary guidelines to reflect this research, backpedaling away from endorsing low-fat, low-cholesterol diets and moving closer to my way of eating.

Another new theme in this book that I haven't covered previously is weight loss. Although I didn't strongly promise weight loss before, I know from thousands of people who took the tenets of *Grain Brain* and *Brain Maker* to heart that weight loss is one of the most common, immediate outcomes of the program. And the weight loss can be huge. You won't feel like you're on a dreaded diet, you won't feel an insatiable hunger, but the pounds will melt away.

I was motivated to write this book from my own personal experience, too. I have tried my best to do everything possible to remain healthy. But now in my sixties, I've experienced my own health issues, and I have learned how to navigate through them successfully within the context of my own principles. I started to think about this book as an opportunity to get myself into the very best shape for the next forty years. Like anyone my age, I'm certainly at risk for all the common disorders. And, by virtue of my family history, I have a higher risk for Alzheimer's disease. But I know I am reducing my risk and stacking the deck in my favor by following the strategies presented in these pages. I want to show you what I have learned and what I do day in and day out.

Some of you are coming to this with the assumption that it is just another diet and lifestyle book that will test your willpower and resolve for a finite period. I'm delighted to disappoint you on that front. *The Grain Brain Whole Life Plan* gives you a jump-start to a healthy way of living that you will be able to sustain indefinitely.

Food is a central component of the program, but so are other key aspects to achieving the best results: the timing of *when* you eat, sleep, and exercise; skipping breakfast once or twice a week; knowing which supplements to take and which medications to potentially drop; reducing daily stress and even chemical exposures in your physical environment; nurturing relationships and your own self-care; addressing the challenges in life with grace and ease; routinely creating goals for your personal development; and finding time for the kinds of physical activities that power the brain while healing the body.

Part I explains the what, why, and how of the program. I'll detail the ground rules, present new data, and offer a 3-step framework that will help you execute my recommendations. You'll start, however, with a prelude to Step 1, during which you will perform a self-assessment to gauge your risk factors, undergo some laboratory tests, and prepare your mind. The main steps are as follows:

Step 1: **Edit** your diet and pill-popping.
Step 2: **Add** your support strategies.
Step 3: **Plan** accordingly.

Part II gives you all the information you need to follow my program, from which foods to eat to which supplements to take and how to leverage the power of sleep, physical movement, and other stress-reducing strategies that will enhance your success.

In Part III, you'll find final tips and reminders, a menu of snack ideas, a basic shopping list, the 14-day meal plan, and delicious recipes to enjoy on your journey. For ongoing support and additional resources, go to www.DrPerlmutter.com.

ON A PERSONAL NOTE

Before we jump into the science in the coming pages, I'd like to share something personal. A lot has happened to me since *Grain Brain* was first published in 2013. In 2015 I lost my dear father, once a brilliant neurosurgeon, to Alzheimer's disease. I also closed my medical practice and took to spreading my message through teaching, the media, and the lecture circuit. I've had the privilege of collaborating with the world's top experts in various fields of clinical medicine and biomedical research whose work further reinforces my recommendations (You'll be reading about some of these people in the book; for videos of my interviews with many of them, please visit www.DrPerlmutter.com/learn).

In early 2016, I came to grips with the sudden, tragic loss of a beloved friend. This was followed by a medical crisis of my own as I lay in the intensive care unit. You'll read about this event later in this book, but suffice it to say that it radically changed my perspective. It vividly taught me the perils of stress and the power of love. And it reinforced the notion that having a healthy mind and body goes far beyond what we eat and how much we exercise.

The day after I was discharged from the hospital, I went to a yoga class with my wife and her mother. At the end of the class, the instructor read a moving passage that struck me right away. It came from the book *How Yoga Works* and basically said that in order for us to reach our highest goals in life, we should try to maintain "...a constant, modest, joyful state of mind which is always looking for ways to protect others from harm—all day long, just in the little world we live in."

Although I am no longer directly involved in patient care on a daily basis, in moving forward, I believe my purpose will be to do just that—to continue to write, lecture, teach, learn, and do my best to protect you from harm. I will keep connecting with people, hearing their stories of transformation, and cheering them on. It's incredibly gratifying to know that I can change people's lives for the better—no surgery or prescription required. May you, too, be a changed person through the execution of a few practical strategies. By reading this book, you are already getting a head start on the path to a better, healthier future.

So no matter what brings you to this book, whether you are concerned for your own health or that of a loved one, rest assured that you have an amazing opportunity right in front of you. And despite your trepidations, this isn't that hard. You've done more difficult things in your life, for sure. Maybe you've given birth, raised a child, cared for someone with special needs, run a company, eulogized a loved one, or battled a serious illness such as cancer. Just getting through life's day-to-day battles is challenging enough. So pat yourself on the back because you've gotten this far, and know that what lies ahead can positively and profoundly change your life for the better.

All that I ask of you at this point is to accept the commitment. You will change your relationship with lots of things in your life, from food to people. You will create new habits and traditions. You will transform how you live your life and reap the ultimate rewards: reaching all those goals I listed earlier. You will not be counting down the days of my 14-day meal plan, waiting for it to end, or suddenly feel like you're force-feeding yourself foods you can't stand no matter how they are prepared. Much to the contrary, you will proceed at your own pace and learn a new way of life that's doable and sustainable *for you* by making a few adjustments to your daily habits.

Take it one day, one new habit, at a time. Be patient and kind with yourself. I have a friend who is also a doctor, and he likes to ask his patients this question: "Who is the most important person in the world?" If they don't answer with a resounding "I am," he teaches them that lesson. Because that's the truth: You are the most important person in the world. Admit that. Live up to that. You deserve it. Choose health. That's the first step to take on the path to radiant wellness.

Welcome to *The Grain Brain Whole Life Plan*. Now let's get going.

PART I

WELCOME TO THE GRAIN BRAIN WHOLE LIFE PLAN

I was given Grain Brain *and* Brain Maker *as birthday presents on January 22, 2016, when I reached age seventy-one, and started a gluten-free, sugar-free, high-fat diet on February 1. After twenty-five days, I have solved two of the three "neurological" health problems I had: left arm trembling while leaning [it] on the arm of a chair, loss of equilibrium, and memory deterioration. The first two are now gone, but for the memory recovery I cannot yet claim an improvement, but I am hopeful. Further, I would like to think that maybe I have improved my speech as well, since before the diet I was reaching the point where it was difficult to have a fluent conversation, because my brain and mouth simply could not speak to each other. I have also lost 3 kilograms [about 6½ pounds] of weight!* —Antonio L.

CHAPTER 1

What Is the Grain Brain Whole Life Plan?

IN THE NEXT EIGHTEEN MINUTES, ***four Americans will die from the food they eat.*** That's one person every four-and-a-half minutes, a fact that's almost impossible to comprehend. But it's heartbreakingly true. That statement was how celebrity chef Jamie Oliver opened his eighteen-minute TED talk a few years ago, stunning the audience and the millions of people who have since watched his video. Oliver has been leading a crusade against the use of processed foods in schools, and he is a staunch advocate for children's rights to wholesome, healthy fare that won't lead to a lifetime of chronic conditions, pain, and illness. It has been postulated that today's generation of children may not live to as ripe an old age as their parents, largely due to the downstream effects of obesity.

But it's not just about children. In developed Western nations, diet-related diseases kill more people than accidents, murder, terrorism, war, *and all other diseases (not diet-related) combined.* Overweight, obesity, type 2 diabetes, high blood pressure, heart disease, dental diseases, strokes, osteoporosis, dementia, and many types of cancer can all be linked somehow to diet. Some of these afflictions have been around for centuries, but not in such epidemic proportions.

I decided to be a neurologist – a doctor who specializes in brain disorders – more than thirty-five years ago. In the early years of my work, I practiced under the idea of "diagnose and adios" for the most part. In other words, once I made a diagnosis I often found that I couldn't offer much to my patients in terms of a treatment, much less a cure. There wasn't anything available at the time, and that was immensely disappointing both for me as a physician and them as patients. I am here to tell you, however, that a lot has changed since then. But it's not all positive. Let me first put a few more facts into perspective.

As you may be aware, over the past century science has made great progress in many areas of medicine. One hundred years ago, the top three causes of death came from infectious germs: pneumonia and flu, tuberculosis, and gastrointestinal infections. Today few of us die from contagions; the top causes of death are noncommunicable illnesses that are *largely preventable*: cerebrovascular disease, heart disease, and cancer. Sadly, while we've made some strides in reducing the rates of some of these chronic illnesses thanks to better prevention and pharmaceuticals, not much has revolutionized my field: averting and treating brain disorders. And these present some of the biggest challenges in medicine. Throughout my career, there have been so many times when I've had to tell patients that I have nothing left in my arsenal to treat them—they have a grave neurological disease that will likely shatter their life and the lives of those they love.

Despite billions of dollars of research, we've had no meaningful treatments or cures for conditions like Alzheimer's disease, Parkinson's disease, depression, ADHD, autism, multiple sclerosis, and so many others. Even chronic conditions like obesity and diabetes, which affect tens of millions today and which are indeed connected to brain disorders, don't have reliable therapies and remedies. A whopping one in five deaths in America is now attributed to obesity, which is among the biggest risk factors for brain-related ailments. You might be surprised to learn that obesity is really a form of malnutrition. As counterintuitive as it sounds, people are overfed and undernourished.

The United States is among the ten wealthiest Western nations where death from brain disease – most commonly dementia – has skyrocketed over the past twenty years. In fact, it leads the way. Since 1979, deaths in America due to brain disease increased an astounding 66 percent in men and 92 percent in women. In America today, it's estimated that 5.4 million people are living with Alzheimer's disease, and that number is predicted to *double* by the year 2030! Someone in the United States develops the disease every 66 seconds; it kills more than breast and prostate cancer combined.

More than 26 percent of adults in the United States—that's about one in four people—suffer from a diagnosable mental illness, from anxiety and mood issues to psychotic disorders, bipolar disorder, and full-blown depression, which is now a leading cause of disability worldwide. One in four women in their prime takes an antidepressant, and may end up staying on that medication for the rest of her life.

When was the last time you had a headache? Yesterday? Right now? Headaches are among the most common brain ailments, and by some estimates they are the number one ailment. More people complain about headaches than any other medical problem. Although nearly everyone has a headache occasionally, one in twenty people has a headache every single day. And an incredible 10 percent of Americans suffer from debilitating migraine headaches—more than diabetes and asthma combined.

Multiple sclerosis (MS), a debilitating autoimmune disease that disrupts the brain's and spinal cord's ability to communicate, affects an estimated 2.5 million people worldwide. Nearly half a million of those patients are in America. The average lifetime cost of treating someone with MS exceeds $1.2 million, and mainstream medicine tells us that there is no cure in sight. Aside from MS, autoimmune disorders in general have been on the rise. I find it interesting and quite telling that, according to people who study ancient diseases, or paleopathologists, humans did not suffer from many autoimmune disorders before the adoption of an agricultural way of life. Autoimmune disease was not nearly as pervasive in the population as it is today. Some autoimmune diseases are three times more common now than they were several decades ago—especially in developed countries like the United States. I love how Lierre Keith, author of *The Vegetarian Myth*, explains it: "That's because it's grains that can turn the body against itself. Agriculture has devoured us as surely as it has devoured the world."

Attention deficit hyperactivity disorder, also known simply as ADHD, has been diagnosed in more than 4 percent of American adults and well over 6 million American children, and an astounding two-thirds of these children are taking mind-altering medications, the long-term consequences of which have never been studied. In fact, 85 percent of the ADHD medicines used across the entire planet are used in America. This is certainly not something to be proud of. Are Americans genetically different from the rest of the world? Or is there something else going on that may be responsible for their overzealous use of drugs?

We also can't ignore the rising prevalence of autism. One in forty-five children age three through seventeen has been diagnosed with autism spectrum disorder (ASD). In boys, ASD is about 4.5 times more common than in girls. The surge in the number of diagnosed cases over the past fifteen years has led some experts to call it a modern-day epidemic. What is going on?

Why have we experienced such a disturbing spike in these afflictions over the past few decades? Why the lack of cures and better treatments? How can only one in one hundred of us get through life without a mental impairment, let alone a headache or two? With so many scientists and so much funding, why has there been so little progress? The answer may simply be that we have been looking in the wrong place. The solution to these challenging disorders may well lie *outside* the brain, and even *outside the body*:

It's in our food.
It's in our gut.
It's in how we live each day and deal with our commitments and responsibilities.
It's in how we move our bodies and stay active, strong, mobile, flexible, and agile.
It's in how we deal with setbacks, illness, injury, and pain.
It's in our relationships and social engagements.
It's in our outlook on life.
And it's in this book.

The Grain Brain Whole Life Plan gives you a way to take control of your mind, body, and spirit. It's a solution to these challenging health problems. It's a way of life. I should emphasize from the get-go that it doesn't just address brain disorders. As I've detailed in previous works, virtually every noncommunicable disease has a lot in common. So whether we're talking about asthma or Alzheimer's disease, diabetes or depression, you might be surprised to learn about the connections among them. You'll read about that soon.

Now, let me play devil's advocate for a moment. Despite our vast knowledge in medicine today, especially compared to what we knew a century ago, the development of disease within the context of the human body still remains puzzling—even to the most educated, brilliant individuals who stay on top of the scientific literature. There's a lot we've figured out: We've cracked the human genome code, our DNA; we've developed advanced diagnostic tools and revolutionized treatments; and we've produced vaccines, antibiotics, and other antidotes to combat known invaders. But in the face of all this, we can struggle mightily to understand why one person dies relatively young while another lives vigorously past ninety. Or why one individual looks 65 at the age of 85, and another appears to be in her 50s when she's really barely 40. We've all heard of the athlete with no documented risk factors for coronary artery disease who dies suddenly of a heart attack; the lung cancer victim who never smoked; and the slim health-nut who is diagnosed with diabetes or early onset dementia. What explains these phenomena?

We have to accept a certain mystery surrounding the body's functionality and whether or not it becomes sick and enfeebled. We also have to acknowledge that how we choose to live—and think—has a significant effect on our health and psychology. It's far easier, and cheaper, to prevent illness than to treat it once it's established. But there is no such thing as "spot prevention" targeting one specific area; we have to honor the body as a whole, complex unit. That is the main idea underlying this program.

Every day I meet people who've tried everything they can to achieve the health that they want and deserve. These individuals often fall victim to dubious, unproven health practices and poor nutrition, and they don't even know it. They complain of various symptoms that share common themes: low energy, difficulty losing weight, digestive disorders, insomnia, headaches, low libido, depression, anxiety, memory problems, burnout, sore joints, relentless allergies. *The Grain Brain Whole Life Plan* is a rallying cry for anyone who hasn't been able to discover true health and maintain that health indefinitely. All roads to perfect health—and ideal weight—begin with simple lifestyle choices.

As I always say, food is more than fuel for the body to survive. Food is information; what I mean by that is that it ultimately has the power to influence how your personal genome—your DNA—expresses itself. In biology this phenomenon is called epigenetics, a concept we'll be exploring shortly. Epigenetics has transformed the way we think about DNA, as well as about food. On a more basic level, food also helps generate the connection between your mind-set and how you *feel.* What you eat directly impacts how you experience life and nourish your body's needs. What you *do*—in your work, in your environment, in your routines, and in your efforts to reduce stress, manage chronic conditions, and address challenges—also affects your body and whether or not you put yourself in harm's way and at risk for serious health issues. And optimizing your body's innate requirements, my friends, is the essence of *The Grain Brain Whole Life Plan.*

The Grain Brain Whole Life Plan can help all of the following:

- ADHD
- asthma
- autism
- allergies and food sensitivities
- chronic fatigue
- chronic pain
- mood disorders, including depression and anxiety
- diabetes and irrepressible cravings for sugar and carbohydrates
- heartburn and gastroesophageal reflux disease, or GERD
- overweight and obesity, as well as weight-loss struggles
- memory problems and poor concentration
- headaches and migraines
- chronic constipation or diarrhoea
- frequent colds or infections
- intestinal disorders, including celiac disease, irritable bowel syndrome, ulcerative colitis, and Crohn's disease

- thyroid dysfunction
- multiple sclerosis
- fibromyalgia
- infertility
- insomnia
- joint pain and arthritis
- high blood pressure
- atherosclerosis
- chronic yeast problems
- skin problems such as acne, eczema, and psoriasis
- bad breath, gum disease, and dental problems
- Tourette's syndrome
- extreme menstrual and menopausal symptoms
- and many more

You don't have to be sick to reap enormous rewards from the plan. Even if you feel relatively good and healthy, you can benefit. So whether you're desperate for a better body and clearer mind or just want to know you're doing all you can to live a healthier, longer life, this program is for you.

Most of you should start to feel the effects of the program within a matter of days. But it will take a little longer for it to have a lasting impact on your body at both the cellular and metabolic levels. It will also take a while to reset your attitude so that you can effortlessly enjoy your new lifestyle. It doesn't matter how often you've failed to follow protocols in the past or how much doubt you have in the effectiveness of my recommendations. What matters is that you focus on your goals and have faith that health and happiness await you.

CHAPTER 2

The Chief Goals

If you're like most people, you can't take time away from your busy life to check into a wellness retreat center or medical spa oasis for a month to concentrate squarely on good nutrition, stress relief, and twice-daily exercise classes worthy of a *Biggest Loser* episode. I've created this book to give you the tools you need to experience maximum results in the shortest period of time. I expect you to continue to go about your daily routine and do your best to make the modifications to your lifestyle that I describe. I will be asking you to start an exercise routine (see page 120) and to seriously consider all the advice I've outlined throughout the book. Some of the strategies will be easy to implement, such as drinking more water throughout the day and keeping a journal to practice gratitude. But some, such as being strict with your sleep schedule, establishing a strength-training routine, spending distraction-free time for self-reflection, and evicting gluten, grains, and sugar from your diet, will likely take time to master. And that's okay. I've included plenty of ideas to help make these strategies doable and practical in today's world.

Unfortunately, many of us live in a highly reactive rather than proactive prevention mode. We avoid taking proper care of ourselves as we chase everything else in life and tend to other responsibilities—and other people. Some of us don't slow down or change our ways until illness or injury strike, and then we are forced to take a detour, if we can find one. We perpetuate negative thinking or say self-sabotaging things to ourselves like "once I achieve X" or "when I make Y dollars, I'll be able to take better care of myself." But as you probably know, this rarely happens in the real world. By the time we're compelled to change, it can be very challenging to do so successfully. And well-meaning intentions to regain health when we could have avoided the problem in the first place don't usually work out the way we want. We can get so run-down and burned out that there's no motivation to do anything but await a serious diagnosis and then come to rely on pharmaceuticals forever. I encounter many people who reach midlife encumbered with chronic conditions or serious illnesses that are not easily treatable or reversible, if at all. Although they may finally have all the resources they need to access high-quality health care, it might be too late. My goal for you is to commence change today to prevent such a fate and relieve any health challenges you currently have so you can enjoy a higher quality of life from here on out. Wouldn't it be wonderful to rely less on drugs and more on your body's natural machinery?

It's amazing to me that in the wake of this epidemic of chronic illness and brain disorders, so few of us stop to think about how our daily lifestyle choices factor into our well-being. It's human nature to prefer the shortcut and ask for a prescription or seek a potion that we believe will make our problems disappear. Yes, there's work and effort involved in choosing to eat a certain way and avoiding the habits that get us into trouble, but it doesn't have to feel like a Herculean task. No sooner do you begin to feel better than you have more motivation to keep going.

So with that in mind, let's take a tour of the chief goals of the program:

- to reduce and control inflammation
- to turn your body into a fat-burning machine using fat
- to balance levels of beneficial bacteria in your belly
- to balance your hormones and increase leptin sensitivity
- to take control of your own genes
- to balance your life

Let's look at each of these goals in turn. I'll remind you of some of the fundamental science as we go.

REDUCE AND CONTROL INFLAMMATION

One of the most paradigm-shifting discoveries in Western science during my career has been that the cornerstone of most diseases and degenerative conditions, including being overweight and being at risk of brain dysfunction, is inflammation. And by now you probably have a rough idea of what "inflammation" means in terms of the body. It's the body's natural healing process whereby it temporarily amps up the immune system to deal with what it thinks is an insult or injury. Whether you're fighting a cold or dealing with a torn muscle, inflammation lies at the heart of your recovery.

The problem with inflammation is that it can become chronic. A water hose turned on momentarily to extinguish a small fire is one thing; but leave the hose on indefinitely, and you've got another problem on your hands. Millions of people are besieged by an inflammatory process that's always in the "on" mode. Their immune systems have been permanently keyed up, but they won't necessarily feel it as they would if they had a laceration or sore throat. This type of inflammation is systemic—it's a slow-boil full-body disturbance that is usually not confined to one particular area. The bloodstream allows it to spread to every part of the body; hence, we have the ability to detect this kind of pervasive inflammation through blood tests.

Many of the biological substances produced as a result of inflammation are injurious to cells, leading to cellular dysfunction and destruction. It's no wonder that the leading scientific research shows that chronic systemic inflammation is a fundamental cause of the morbidity and mortality associated with all manner of disease and virtually every chronic condition you can imagine. Even your mood is affected by inflammation. One of the first things I hear from people I put on my protocol is that it has more than a physiological impact; it has a tremendous psychological effect, too. And new science tells us that mood disorders as severe as depression are, in fact, rooted in inflammation—not necessarily in low or misbehaving brain chemicals.

The Grain Brain Whole Life Plan turns on the pathways in your body that help reduce and control inflammation. You'll be embracing an anti-inflammatory lifestyle and applying basic strategies to your everyday habits that are designed to lower inflammation. Natural substances like those you'll find on this diet (for example, turmeric) have been described in medical literature for more than two thousand years, but it is only in the past decade that we have begun to understand their intricate and eloquent biochemistry. And it's not just what you eat that can help you manage inflammation. You're going to learn about the latest studies on how exercise and sleep come into play, too.

TURN YOUR BODY INTO A FAT-BURNING MACHINE USING FAT

A central premise of *Grain Brain* is that fat—not carbohydrate—is our metabolism's preferred fuel and has been for all of human evolution. I made my case for choosing high-quality fats and not worrying about so-called "high cholesterol" foods. Nutritional therapist Nora Gedgaudas states it perfectly in her book *Primal Body, Primal Mind*: "99.99% of our genes were formed before the development of agriculture." As *Homo sapiens*, we are virtually identical to every human that has walked on the planet. And, as a species, we have been shaped by nature over thousands of generations.

Throughout human evolution, and for the greater part of the last 2.6 million years, our ancestors' diets consisted of wild game and seasonal fruits and vegetables. We sought fat as a calorie-dense food source. It kept us lean and served us well in our hunter-gatherer days. In fact, we consumed a diet estimated to contain as much as ten times more fat than our current intake. Today most people fear dietary fat, equating the idea of eating fat to *being* fat. The truth is quite different. Obesity and its metabolic repercussions have almost nothing to do with dietary fat consumption and everything to do with an addiction to carbs. People continue to gravitate toward "fat-free," "low-fat," "multigrain," and "whole grain" labels, foods that contain ingredients whose downstream effects assault the brain and body. Eating carbohydrates stimulates insulin production, which leads to fat production, fat retention, and a reduced ability to burn fat (much more on insulin shortly). Dietary fat does not do this. What's more, as we consume carbohydrates, we trigger an enzyme called lipoprotein lipase that tends to drive fat into the cell; the insulin secreted when we consume carbohydrates makes matters worse by triggering enzymes that promote fat storage.

> The human dietary requirement for carbohydrate is virtually *zero*.

When I tell people that we can survive—*thrive*—on zero dietary carbohydrates but lots of dietary fat, cholesterol included, I am sometimes met by bewildered faces. But these days, that's changing. Up until quite recently, we had been told that the brain needs glucose to survive and that we supply that nutrient through carbs. The science has finally prevailed, and now it's clear that, yes, the brain needs glucose, but our bodies can make glucose. I repeat: It is the sugar we consume that makes us fat—not the dietary fat.

The same is true of cholesterol: Eating foods high in cholesterol has no impact on our actual cholesterol levels, and the alleged correlation between higher cholesterol levels and higher cardiac risk is untrue. We have been consuming animal protein and saturated fat for the past 100,000 generations. Yet we've been told that saturated fat is dangerous. The fact that approximately 50 percent of the fat in human breast milk is saturated should go a long way to highlight the value and importance of saturated fat.

So what happens when you substantially reduce your carbohydrate intake and derive more of your calories from fat? You turn your body into a fat-burning machine. When you follow a diet low in carbohydrates, minimal in protein, and rich in healthy fats and plant-based fibers, you stimulate the body to use fat rather than glucose for fuel. More specifically, you force the body to turn to specialized substances called ketones for energy. In the absence of carbohydrates, ketones are produced by the liver using fatty acids from your food or body fat. These ketones are then released into the bloodstream, where they can travel to the brain and other organs to be used as fuel. A so-called ketogenic diet—one that derives 80 to 90 percent of calories from fat, and the rest from fibrous carbohydrates (e.g., whole fruits and vegetables) and high-quality protein—is the foundation of the Grain Brain Whole Life Plan.

There is nothing new or faddish about the ketogenic diet. Versions of the diet have been used for centuries, and it may even date back to biblical times. It has been successfully used to treat drug-resistant epilepsy in children since the 1920s. New evidence is emerging from animal research and clinical trials to show that ketogenic diets help treat an array of neurological disorders, from headaches and sleep disorders to bipolar disorder, autism, brain cancer, and other diseases.

Your body is in a state of ketosis when it's creating ketones for fuel instead of relying on glucose. A mild state of ketosis is healthy. We are mildly ketotic when we first wake up in the morning, as our liver is mobilizing body fat to feed our hungry organs. Both the heart and the brain run more efficiently, by as much as 25 percent, on ketones than on blood sugar. The brain's energy needs account for 20 percent of total energy expenditure, and healthy, normal brain cells thrive when fueled by ketones.

Neurological diseases may all have their distinct characteristics and underlying causes, but one feature they share in common is deficient energy production. When the body uses ketones to maintain normal brain cell metabolism, some of those ketones are a more efficient fuel than glucose, as ketones provide more energy per unit of oxygen used. Being in ketosis also amplifies the number of mitochondria in brain cells, the cells' energy factories. Studies have documented that ketosis shores up the hippocampus, the brain's main center for learning and memory. In age-related brain diseases, hippocampus cells often degenerate, leading to cognitive dysfunction and memory loss. But with greater energy reserves, the neurons are better buffered against disease stressors.

I should add that despite entering ketosis, the body's blood glucose levels remain physiologically normal. You won't experience the ills of low blood sugar because the body can derive glucose from certain amino acids and the breakdown of fatty acids. (And, as we'll see in Part II, there's one ketone in particular that serves as an excellent alternative fuel source that also has the power to prevent the body from breaking down muscle tissue to generate glucose.)

The diet protocol outlined in Part II honors the main ketogenic principles of significantly reducing carbohydrates to the point that the body is pushed to burn fat, while dietary fat and other nutrients turn on the body's powerful "pro-health" technology. The key, of course, is to eat the right kind of fat. I'll explain more shortly.

BALANCE LEVELS OF BENEFICIAL BACTERIA IN YOUR BELLY

During my public talks, I like to refer to "Slick Willie" Sutton, one of the most infamous and prolific bank robbers of the 20th century, who was once asked why he robbed banks. His answer: "Because that's where the money is." Likewise, you'd think that if you wanted to understand problems with the brain, that's where you'd need to look, right? But here's where the story gets interesting. Recent research has shown that the root of many brain-related disorders could be not within the brain but rather within the body, especially within the gut. Let me repeat: What's taking place in your intestines today plays a critical role in determining your risk for any number of brain conditions. Which is why optimizing intestinal health and maintaining the structure and function of the gut barrier—the wall that separates the interior of your gut and your bloodstream—is paramount.

First, a quick anatomy lesson. The gut is, at a very basic level, a biological "pipeline" that goes from the mouth to the anus. Anything you consume that isn't digested will pass through you and go out the other end. One of the most crucial functions of the gut is to prevent foreign substances from entering your bloodstream and reaching vulnerable organs and tissues, including the brain.

The gut and the brain are, in fact, intricately connected. The gut has an impact on the brain's function in the moment as well as in the long term; it influences your risk of developing a neurodegenerative condition, like Alzheimer's or Parkinson's disease. *Brain Maker* covered the science of the microbiome in depth, especially as it relates to brain health, and since its publication newer studies have continued to confirm the facts. For example, in 2015 a landmark European study found a powerful relationship between an unhealthy intestinal microbiome, often referred to as gut dysbiosis, and the development of Parkinson's disease. Some are even calling the intestinal microbiota, or gut flora, the brain's "peacekeeper."

Just what exactly makes up the human microbiome? It consists of a large and extended family of more than 100 trillion organisms—mostly bacteria living within the gut—that outnumber your own cells ten to one. These organisms' metabolic products and genetic material are also considered to be part of the microbiome. Amazingly, a full 99 percent of the genetic material in your body is housed by your microbiome! It supports and nurtures every aspect of your physiology, including what goes on in the brain.

We now know that our lifestyle choices help shape and sustain our microbiome. We also know that the health of the microbiome factors into immune system function, inflammation levels, and risk for illnesses as diverse as depression, obesity, bowel disorders, multiple sclerosis, asthma, and even cancer. Indeed, the National Cancer Institute has recently revealed that certain gut bacteria regulate and "educate" the immune system in such a way that they can help reduce the growth of tumors; what's more, gut bacteria help control the efficacy of certain well-established anticancer therapies. They also do a lot of work on behalf of our physiology: They manufacture neurotransmitters and vitamins that we couldn't otherwise make, promote normal gastrointestinal function, provide protection from infection, regulate metabolism and the absorption of food, and help control blood sugar balance. They even affect whether we are overweight or lean, hungry or satiated.

New science is emerging that demonstrates that microbes not only influence the activity of our DNA, moment to moment, but also have become part of our own DNA throughout evolution. In other words, microbes have inserted their genes into our own genetic code to help us evolve and flourish. Isn't that astonishing? In the words of one prominent team of researchers at Stanford and the University of California, San Francisco: "Recent discoveries make clear that our microbiota is more like an organ than an accessory: These microbes are not just key contributors to human health but a fundamental component of human physiology." The National Institute of Health is investing more than $190 million over a five-year period to look at how the human microbiome influences gene expression. And in May 2016, the US government launched a $500 million project called the Unified Microbiome Initiative to study the microbial communities on Earth. This will be a coordinated collaboration among numerous federal agencies, universities, philanthropic organizations, and industry.

In neurology, a field that has essentially been devoid of real cures and treatments, this burgeoning new area of study is finally giving us revolutionary approaches to relieve suffering. My recommendations in Part II leverage the power of this dazzling new science, showing you how you can take advantage of it for your own health. Best of all, you can reap the rewards within a matter of *days*.

> Microbiomes exist throughout the world. In addition to the human microbiome, the oceans, soil, deserts, forests, and atmospheres each have their own microbiomes that support life. Microbiomes have captivated the scientific world, fueling multimillion-dollar research projects globally. A nifty fact: Microbes in our oceans produce 50 percent of the oxygen we breathe while absorbing carbon dioxide. Methane-ingesting microbes that inhabit the deep sea also act as a powerful incinerator for that notorious greenhouse gas. It's important to recognize that the health of the planet depends on its microbial communities.

One of the key areas that intestinal bacteria help control is gut permeability. When we are talking about permeability issues in the gut, or so-called "leaky gut," we are referring to problems in the competency of the tight junctions—the small connections between the cells that line the gut and control the passage of nutrients from the gut into the body via your circulation. If the junctions are somehow compromised, they fail to appropriately police what should be allowed in (nutrients) or kept out (potential threats). As the gatekeepers, these junctions determine, to a large extent, your body's set point of inflammation—your baseline level of inflammation at any given time.

The term "leaky gut" used to be dismissed by conventional researchers and doctors, especially as it relates to autoimmunity. But now an impressive number of well-designed studies have repeatedly shown that when your intestinal barrier is damaged, which can result in having an unhealthy gut flora that cannot protect the intestinal lining, you are susceptible—through increased inflammation and an activated immune response—to a whole spectrum of health challenges, including rheumatoid arthritis, food allergies, asthma, eczema, psoriasis, inflammatory bowel disease, celiac disease, type 1 and type 2 diabetes, and even cancer, autism, Alzheimer's, and Parkinson's.

According to this new science, the intestinal wall has everything to do with whether we tolerate or adversely react to substances we ingest. A break in that intestinal wall can cause food toxins such as gluten and pathogens to pass through and agitate the immune system. This breach affects not only the gut, but also other organs and tissues, such as bones, skin, kidneys, the pancreas, liver, and brain.

What can cause an unhealthy intestinal microbiome?

- diets high in refined carbohydrates, sugar, and processed foods
- diets low in fiber, especially the kind that feeds the flora
- dietary toxins, such as gluten and processed vegetable oils
- chronic stress
- chronic infections
- antibiotics and other medications like nonsteroidal anti-inflammatories (NSAIDs) and acid-reflux drugs (proton pump inhibitors or PPIs); see page 114

When researchers at Stanford University, led by Dr. Justin Sonnenberg, explored the mucus layer that lines the gut, they found that it is home to several groups of bacteria that are vital for regulating immunity and inflammation. The mucus layer, which renews itself every hour, is critical in maintaining the integrity of the gut lining and reducing leaky gut. It's becoming clear that the bacteria in this layer depend on dietary fiber to thrive, which is why the carbohydrates we do consume should be in the form of fiber-rich fruits and vegetables. These complex carbohydrates are broken down by our gut bacteria. That's right: The gut's beneficial bacteria use the fiber we eat as fuel to promote their own growth.

Prebiotics are a specialized form of dietary fiber that our bodies cannot digest but that gut bacteria love to consume, and they are an important part of the Grain Brain Whole Life Plan. Prebiotics are often categorized as carbohydrates because they are found in many fruits and vegetables. Prebiotics act like a fertilizer; it has been estimated that for every 100 grams of prebiotics consumed, a full 30 grams of bacteria are produced. As our gut bacteria metabolize this fiber, they produce substances called short-chain fatty acids (SCFAs), which help us stay healthy. Butyric acid, for example, is an SCFA that improves the health of the intestinal lining. In addition, these fatty acids help regulate sodium and water absorption and enhance our ability to take in important minerals and calcium. They effectively lower the pH in the gut, which inhibits the growth of potential pathogens and damaging bacteria. They enhance immune function and even help explain why some people have trouble losing weight, even though they cut back on calories. The production of these SCFAs effectively activates a signaling pathway that tells the brain that the body has gotten enough food. This messaging in turn triggers food in the gut to move more quickly, so there is less absorption of calories. On the other hand, when SCFAs are low, the body believes it's not getting enough food, and so food moves more slowly, allowing the body to extract more calories.

The typical Western diet supplies plenty of calories but little or no prebiotic fiber. So despite our huge caloric intake, our digestive system believes we are starving! The body reacts to this misguided sensation of starvation by doing what it can to extract as many calories as possible from our food. This may well represent one of the primary issues underlying obesity. The average American consumes a scant 5 grams of prebiotic fiber daily, while estimates reveal that their slender hunter-gatherer forebears may have consumed as much as 120 grams each day. I'll show you how to stock up on prebiotic fiber so your body doesn't believe it's in starvation mode and doesn't have to forage as many calories from the food you eat.

New research also reveals that gut bacteria play an important role in maintaining the blood-brain barrier, which protects the brain from potentially harmful substances. The blood-brain barrier ensures homeostasis of the central nervous system, too. In fact, there are many newly discovered similarities between the blood-brain barrier and the gut's lining. It was recently demonstrated, for instance, that gliadin, a protein found in gluten, may lead to increased permeability of the blood-brain barrier, just as it leads to increased permeability of the gut. This could further explain the relationship between gluten-containing foods and neurological problems. So if you thought having a leaky gut was bad, just imagine what might happen with a leaky brain! As a matter of fact, problems with the blood-brain barrier have been associated with Alzheimer's disease, stroke, brain tumors, multiple sclerosis, meningitis, rabies, seizures, and even autism.

In the fall of 2014, I had the opportunity to speak at Harvard Medical School on the role of the microbiome in brain health and disease. Just before it was my turn to speak, I had a chat with my friend and colleague Dr. Alessio Fasano, one of the world's leading authorities on gluten and health, who was also presenting. Dr. Fasano, who heads the Center for Celiac Research at Harvard's Massachusetts General Hospital, made it clear that, in his opinion, the number one factor shaping the microbiome is diet. And we have control over our diet.

Dr. Lawrence David, another Harvard researcher, has addressed the question of how long it takes for the human microbiome to change once the diet has changed. His study, published in January 2014, assessed changes in the gut bacteria that occurred in six men and four women between the ages of twenty-one and thirty-three when they consumed either an animal-based diet or a plant-based diet. Although it was a small study involving only a handful of people, it nonetheless spurred further research. Dr. David documented fairly dramatic changes in the genetic signature of the gut bacteria in as little as *three days*. In another, collaborative study involving a consortium of researchers from Germany, Italy, Sweden, Finland, and the United Kingdom, it was found that "A key factor in determining gut microbiota composition is diet... Western diets result in significantly different microbiota compositions than traditional diets." Their investigations have also found that the way the gut bacteria function can vary greatly depending on diet. And they noted differences in gene expression of the bacteria, too.

My bet is that the science will increasingly highlight the power of "traditional diets," high in healthy fats and low in carbohydrates, and the perils of the Western diet, high in carbs and low in healthy fats.

It may have been bold to write a whole chapter in *Brain Maker* about the connection between gut health and risk for autism, but the science continues to confirm this link. A pattern of gastrointestinal disorders, likely brought on by leaky gut and gut dysbiosis, are routinely documented in children with autism. One theory currently making waves supposes that distinctive microbes associated with autism spectrum disorder create by-products in the microbes' metabolism that in turn affect human brain function. So powerful is the science that the FDA has approved a study at Arizona State University in which doctors will perform a fecal microbial transplant (FMT) in a group of twenty children aged seven to seventeen years with autism who also have severe gastrointestinal problems. FMT is the most aggressive therapy available to reset and recolonize a sick microbiome. In the procedure, filtered good bacteria is transplanted from a healthy person into the colon of another person.

The incredible role of the microbiome in keeping you healthy is captivating researchers around the globe. You'll be learning more throughout the book, including what you can do starting today to balance your gut's microbial community and avoid gut dysbiosis. For now, however, let's briefly turn to another chief goal of this program.

BALANCE YOUR HORMONES, REDUCE INSULIN SURGES, AND INCREASE LEPTIN SENSITIVITY

Your endocrine system, which manages and controls your body's hormones, holds the remote control to much of what you feel—moody, tired, hungry, sexual, sick, healthy, hot, or cold. It lords over development, growth, reproduction, and behavior through an intricate system of hormones, the body's chemical messengers. These messengers are manufactured in different parts of the body (for example, the thyroid, adrenal, or pituitary gland or gonads) and then travel through the bloodstream to reach target organs and tissues. Once there, they act on receptors to elicit a biological response—usually with the goal of effecting change that allows your body to run smoothly and maintain balance. They serve a vital role in every bodily system, including your reproductive, nervous, respiratory, cardiovascular, skeletal, muscular, immune, urinary, and digestive systems. To keep the body balanced, the forces of one particular hormone are usually counterbalanced by those of another hormone.

Hormone imbalances can lead to serious health issues—metabolic and thyroid disorders, infertility, cancer, hair loss, fatigue, depression, a loss of libido, chronic pain, and more. Hormonal chaos can happen quite naturally during stressful periods or as a result of your age or various health conditions that disrupt the harmony. Women experience decreases in oestrogen and fluctuations in thyroid hormones during and after menopause, while men have a drop in their testosterone levels by 1 to 2 percent every year after age thirty (part of this drop, however, can be attributed to lifestyle factors—typically weight gain—rather than aging alone). As we've just seen, a misbehaving microbiome can certainly come into play as well. And hormone levels can be affected by certain toxins.

The good news is that hormonal dysfunction can often be addressed through diet...and through the plan of action outlined in this book. One key factor is probiotics ("for life"), live bacteria you can ingest through foods and supplements. Remarkable new studies have emerged that show just how powerful probiotics can be in balancing insulin, the body's master hormone, as well as other hormones related to appetite and metabolism; they can help reduce and even eliminate insulin resistance and diabetes.

Let me give you a quick primer on insulin and some of the other important hormones related to metabolism. Insulin, as you likely already know, is one of the body's most important hormones. A carrier protein produced by the pancreas, insulin is best known for helping us transport carbohydrate-based energy in the form of glucose from food into cells for their use. Insulin circulates in your bloodstream, where it picks up glucose and moves it into cells throughout the body, where it can then be used as fuel. Extra glucose that the cells don't need is stored in the liver as glycogen or deposited in fat cells.

Normal, healthy cells have no problem responding to insulin. But when cells are relentlessly exposed to high levels of insulin as a result of persistent spikes in glucose—again, typically caused by consuming too many modern carbohydrates—our cells adapt and become "resistant" to the hormone. This triggers the pancreas to pump out more, so now higher levels of insulin are required for glucose to enter cells. But these higher levels also cause blood sugar to plummet to dangerously low levels, resulting in physical discomfort and brain-based panic.

As I detailed in *Grain Brain*, the connections among high blood sugar, insulin resistance, diabetes, obesity, and risk for brain disorders are irrefutable. Studies show not only that high body fat correlates with a smaller hippocampus, the brain's memory center, but also that the metabolic consequences of diabetes and obesity have far-reaching effects on the brain. After 1994, when the American Diabetes Association recommended that Americans should consume 60 to 70 percent of their calories from carbohydrates, the rate of diabetes exploded. And so did the rate of brain disorders. People with diabetes have twice the risk of Alzheimer's disease.

The exact nature of that relationship was just recently brought to light. For starters, if you're a diabetic, by definition you have high blood sugar because your body cannot transport critical glucose into cells. And if that glucose remains in the blood, it will inflict a lot of damage. It will attach to proteins in the body in a process called glycation, which then triggers inflammation as well as the production of free radicals. All of these—glycation, inflammation, and free radical production—are implicated in Alzheimer's, Parkinson's, and multiple sclerosis. Even being prediabetic, when blood sugar issues are just starting to arise, is associated with a decline in brain function and a risk factor for full-blown Alzheimer's disease.

In 2016, Melissa Schilling, a professor of management and organizations at New York University, added to our understanding of how diabetes and Alzheimer's disease are related when she uncovered a pathway between these two diseases. Integrating decades of research on molecular chemistry, diabetes, and Alzheimer's, she found a commonality: insulin and the enzymes that break down this important hormone. The same enzymes that break down insulin also break down amyloid-beta, the protein that forms tangles and plaques in the brains of people with Alzheimer's. When people secrete too much insulin due to a poor diet, obesity, and diabetes (a condition called hyperinsulinemia), the enzymes are too busy breaking down insulin to break down amyloid-beta, causing amyloid-beta to accumulate. Schilling's work has led to a stunning new fact: Almost half of all Alzheimer's disease cases in the United States are likely due to hyperinsulinemia. Fortunately, hyperinsulinemia is preventable and treatable using this very protocol. (I recently had the opportunity to interview Professor Schilling for *The Empowering Neurologist* online program, and you can watch this compelling video on my website, www.DrPerlmutter.com.)

There are two other important hormones related to metabolism that share a relationship with insulin: leptin and ghrelin. The biochemistry of all three hormones in the body is intricately complex and a highly regulated affair, but I'm going to distill it down for you so you understand why it's so important to keep these critical hormones balanced.

Leptin and ghrelin are your two chief appetite hormones. Whereas insulin controls energy use and storage upon the intake of food, leptin and ghrelin control whether you feel hungry or full—they orchestrate the stop and go of our eating patterns. Leptin, from the Greek word for "thin," is involved in dozens of bodily processes, including helping to coordinate the body's inflammatory responses, but it is best known for its role in appetite suppression. It reduces the urge to eat by acting on specific centers of the brain. As nutritional therapist Nora Gedgaudas is fond of saying, leptin tells your brain that "the hunting is good." It's what allows you to put that fork down and stop eating. Here's how it works in a nutshell: When fat cells start to fill up and expand, they secrete leptin. Once the fat cells begin to shrink as their contents are burned for energy, the tap is slowly turned off and less leptin gets released. Eventually you're able to feel hunger again, thanks to the release of ghrelin, and the cycle starts all over.

Ghrelin, the "hunger hormone," is triggered by an empty stomach and increases your appetite. As the stomach fills with food and expands, signals to your brain tell the ghrelin tap to turn off. As you can imagine, a disruption in the balance between leptin and ghrelin will wage war on your cravings, sense of fullness, and waistline. People who are leptin resistant don't feel full (and can't stop eating). Gedgaudas calls leptin resistance the Holy Grail of obesity. In the same way that too much insulin pushes you toward insulin resistance (and diabetes), too much leptin, triggered by an overload of dietary carbs and sugar, leads to leptin resistance. And high levels of insulin render the brain less sensitive to leptin.

Due to the prevalence of insulin resistance today, most people—regardless of their weight—release twice as much insulin as people did just 30 years ago for the same amount of glucose. And that high insulin is responsible for perhaps 75 to 80 percent of all obesity.

Obviously, the goal is to not only achieve optimal blood sugar control through healthy insulin levels, but also to balance the relationship between leptin and ghrelin, and, in particular, to increase your body's sensitivity to leptin. I'm going to show you how to do that not just through diet but also through sleep and exercise. Sleep deprivation reduces leptin, so your brain gets the message to seek out more calories; exercise improves leptin signaling (as well as insulin sensitivity).

TAKE CONTROL OF YOUR OWN GENES

When you think of your DNA, your inherited genetic code, you probably think about what kinds of characteristics and risk factors your biological parents bestowed on you through their own DNA. Did they give you blue eyes, an athletic build, and a propensity to have heart trouble later in life? We used to think that DNA was like a permanent marker in your body's chromosomes. You couldn't change it. But now we know that even though genes encoded by DNA are essentially static (barring the occurrence of mutation), the *expression* of those genes can be highly dynamic.

I mentioned one of the hottest areas of research earlier: epigenetics, the study of sections of your DNA (called "marks" or "markers") that influence how your genes act and behave. Put simply, these epigenetic markers have a say in your health and your longevity, as well as the health and longevity of your own children. Indeed, the forces acting on the activity of your DNA today—for good or bad—can be passed on to your future biological children. Epigenetic activity may even change your *grandchildren's* risks for certain diseases and disorders. By the same token, these markers can be changed to affect your DNA's expression differently, making it fully possible to change your underlying risk for certain diseases.

Epigenetic forces can affect us from our days in utero to the day we die. There are many windows during our lifetime when we are extra sensitive to environmental influences that can change our biology and have downstream effects such as dementia and brain cancer.

There's one important molecule I'd like to highlight that has everything to do with your ability to control your own genes' expression: Nrf2. When the body experiences high oxidative stress, which is another way of saying there's an imbalance between the production of free radicals and the ability of the body's ability to counteract their harmful effects, it sounds the alarm by activating Nrf2, a specific protein found within every cell. This protein remains dormant, unable to move or operate, until it is released by an Nrf2 activator. Once activated, it then migrates into the cell nucleus and bonds to the DNA at a specific spot, which then opens the door for the production of a vast array of important antioxidants as well as detoxification enzymes. The result is both the elimination of harmful toxins and the reduction of inflammation.

The Nrf2 pathway's chief role is to protect cells against external stresses such as toxins and carcinogens. It is an ancient circuit. As described in a 2014 paper by researchers at the University of Colorado, the Nrf2 pathway has been referred to as the "master regulator of antioxidant, detoxification, and cell defense gene expression." For these reasons, a great deal of research has been carried out on the role of this life-sustaining pathway, especially in disorders such as Alzheimer's disease, Parkinson's disease, multiple sclerosis, and even autism.

But you don't need to wait until the body sounds its own alarm to activate the Nrf2 pathway. You can turn it on through the consumption of certain ingredients in the diet and through calorie restriction. The healthy omega-3 fat DHA, found in many fish, acts directly upon the Nrf2 pathway, as do compounds found in broccoli, turmeric, green tea extract, and coffee. You'll find these ingredients recommended in the dietary protocol. And calorie restriction will happen quite naturally due to the nature of the protocol's low-carb approach as well as the occasional fast (more on this soon).

In the past couple of years, scientists have discovered that lactobacilli—good bacteria that are very much a part of the gut's community and can be found in probiotics—stimulate the Nrf2 pathway. In experimental studies, these good bacteria allow animals to respond to stress by turning on their protective genes through the Nrf2 pathway. This illustrates the true power of our friendly gut bacteria. Not only are they participating in creating vital substances we need to survive, but they also are creating an environment that influences the expression of our genes for the better.

I've never included a discussion about telomeres in my books before, but the science is finally giving us clues to how important they are, as well as to what can adversely affect them. Telomeres are the caps on the ends of the chromosomes. Because they protect our genes and make it possible for cells to divide, they are critical to our health and are believed to hold secrets on how we age and develop disease. In terms of brain disease, for example, it has recently been shown by researchers at the Karolinska Institute in Sweden that "telomeres are involved in the actual active mechanism behind the development of [Alzheimer's] disease...."

Oxidative stress, brought on by psychological stress or too much sugar and carbs, has been shown to shorten telomeres and, thus, life. The shorter the telomeres, the faster we age. Smoking, exposure to pollutants, and obesity also cause oxidative stress, thereby shortening telomeres. On the other hand, we can protect our telomeres with aerobic exercise, reduced sugar, more dietary fiber, and added DHA. This protocol will help you do just that.

BALANCE YOUR LIFE

We all want it: more balance in our lives. More harmony between work and play, and more strength to overcome difficulties, especially when they are unexpected. We all have things that can derail us, whether physical challenges or mental ones. I trust that once you implement the strategies in this book, you'll enjoy an all-around better, more balanced life. Now let's get to the rules.

CHAPTER 3

The Food Rules

YOUR BODY IS AN INCREDIBLY dynamic, self-controlled machine. It has built-in systems of checks and balances to keep it on an even keel. Just because you pig out and shun exercise one day, for example, doesn't mean you'll gain ten pounds overnight. The body doesn't work like that. Cellular transactions are happening every second, without your even knowing it, that help to maintain your body's overall balance and preferred settings—what we call homeostasis. Consider your personality as an analogy. It remains relatively constant even though you have good days and bad days, times when your mood dips and others when you feel elated.

The body changes day by day based on your experiences and how you treat it, yet it tends to have a general baseline—a state of being where hormones and other biomolecules are flowing as they should, neurons are firing properly, and your immune system is working for you, not against you. Trouble, however, can arise from overriding the body's systems that maintain that homeostasis. Suddenly we can find ourselves vulnerable to illness, disorder, and disease. Nowhere is it easier to open the door to dysfunction than through the assaults we inflict on our body from our daily dietary choices.

I recommend that you keep a daily journal to record what is happening in your life as you move forward. You can write down not only your reasons for pursuing this new way of life, but also your thoughts, goals, and the events that are most affecting you and how you make decisions. See if you can maintain an ongoing record of your feelings and emotions, especially those around food and eating. Catch yourself when you eat mindlessly because you're tired or stressed out. Find patterns between your emotional well-being and the choices you make in daily life. Your attitude and perspective have a big impact on your daily decisions and overall health, and you can learn to use contentment as well as frustration or disappointment as a motivator on your path to success. On the bad days, which are inevitable, aim to be extra vigilant about how the challenging moments affect those behavioral patterns that prevent you from engaging in healthy activities. Such self-awareness will help you to make positive changes and avoid getting derailed by a vending machine or pushy coworker who brings in a box of jelly doughnuts.

My wish for you is that you learn to live in a way that you can sustain for the long term. At this juncture, all I ask is that you make the most of my recommendations and tune in to how your body is feeling and changing. You are recalibrating one day, one meal, one thought at a time and will see the results build up over time. So take a deep breath, relax, and get ready to discover a whole new you.

Now let's get to the training grounds. Time to learn the dietary rules:

- Evict gluten (even if you don't think you have a problem with it)
- Go low-carb, higher fat and fiber
- Abandon sugar (real, processed, and artificial)
- Avoid GMO foods

- Watch out for too much protein
- Embrace the incredible egg

EVICT GLUTEN (EVEN IF YOU DON'T THINK YOU HAVE A PROBLEM WITH IT)

I wrote extensively about gluten in *Grain Brain*, calling the "sticky" protein found in wheat, barley, and rye among the most inflammatory ingredients of the modern era. I argued that while a small percentage of the population is highly sensitive to gluten and suffers from celiac disease, it's possible for virtually *everyone* to have a negative, albeit undetected, reaction to gluten. And now my position has been validated by many fine research groups, including a consortium of scientists from Harvard University, Johns Hopkins, the Naval Medical Center, and the University of Maryland, who published their findings in 2015. My position may have seemed bold, aggressive, and seemingly outrageous and controversial at the time, but it has been confirmed over and over again in the scientific literature since then. Let me give you more details and provide updated evidence.

Gluten sensitivity—with or without the presence of celiac—drives the production of inflammatory cytokines, which are pivotal players in neurodegenerative conditions. The brain is among the organs most susceptible to the deleterious effects of inflammation. And the downstream inflammatory effects of gluten reach the brain via a leaky gut that fails to prevent the toxic ingredient from igniting an immune response. Gluten is a silent poison because it can inflict lasting damage without your knowing it. Those who experience symptoms of gluten sensitivity complain primarily of abdominal pain, nausea, diarrhoea, constipation, and intestinal distress. They also can suffer neurological symptoms such as headaches, brain fog, feeling unusually tired after a gluten-containing meal, dizziness, and a general feeling of being off balance. Most people, however, have no obvious symptoms yet could be experiencing a silent attack somewhere in the body; for example, in the nervous system. While gluten's effects might start with unexplained headaches, chronic fatigue, and anxiety, they can worsen to more dire disorders, such as depression and dementia. It's important to understand that you don't have to experience gastrointestinal symptoms to have a leaky gut. As I explained in Chapter 2, this condition can manifest as an autoimmune disorder, skin problems such as eczema or psoriasis, heart disease, and the spectrum of brain-based challenges.

Although there used to be a debate about whether someone without celiac disease could be sensitive to gluten, science has spoken. Non-celiac gluten sensitivity (NCGS) is finally a diagnosis in mainstream medicine. In one of the most stunning papers of late, published in *Clinical Gastroenterology and Hepatology*, a group of Italian researchers performed a rigorous study (i.e., randomized, double-blind, placebo-controlled) to determine the effects of giving low doses of gluten to people with suspected NCGS. Participants were randomly assigned to consume a little over 4 grams of either a gluten-containing product (approximately the amount in two slices of wheat bread) or a non-gluten-containing product (rice starch), which acted as the placebo, for one week. During that week, participants didn't know whether they were getting gluten or not. They were then put on a gluten-free diet for one week, and after that, participants switched groups. The researchers found a clear relationship between gluten and intestinal symptoms, irritation around the mouth, and, notably, foggy mind and depression; that is, non-intestinal symptoms. They reported: "We found that the overall symptom score was significantly higher while ingesting gluten in comparison with placebo."

Gluten is everywhere today, despite the gluten-free movement taking hold among food manufacturers. It lurks in everything from wheat products to ice cream to hand cream. It's even used as an additive in seemingly "healthy" wheat-free products. I hear about the effects of gluten every day from people I encounter. Regardless of what ails them, whether chronic headaches, anxiety, or a host of neurological symptoms with no definite diagnosis, one of the first things I do is suggest the total elimination of gluten from their diets. And I continue to be astounded by the results. I don't even recommend gluten sensitivity tests anymore. **You must operate from a place of assuming that you are sensitive to gluten and avoid it entirely.**

It's crucial to understand that gluten is made up of two main groups of proteins, the *glutenins* and the *gliadins*. You can be sensitive to either of these proteins or to one of the twelve smaller units that make up gliadin. A reaction to any of these can lead to inflammation. Gliadin in particular has been implicated in new studies showing the protein's damaging effects on the gut lining, inducing permeability. In the words of Harvard's Dr. Alessio Fasano, "...gliadin exposure induces an increase in intestinal permeability in all individuals, regardless of whether or not they have celiac disease."

In 2015 Dr. Fasano published a landmark paper showing how gliadin can wreak so much havoc and even be the culprit behind autoimmune disorders and cancer. Briefly, gliadin triggers production of another protein called zonulin, which breaks down the gut lining and increases permeability. Once the lining is compromised, as you already know, substances that are supposed to stay in the gut leak into the bloodstream and incite inflammation. The discovery of zonulin's effects on the body inspired researchers to look for illnesses characterized by intestinal permeability. And lo and behold, this led to the finding that most autoimmune disorders, including celiac disease, rheumatoid arthritis, multiple sclerosis, type 1 diabetes, and inflammatory bowel disease, are distinguished by abnormally high levels of zonulin and a leaky gut. So powerful is zonulin that when scientists expose animals to the toxin, the animals develop type 1 diabetes almost immediately; the toxin induces a leaky gut, and the animals start making antibodies to islet cells. These are the cells responsible for making insulin.

For those of you trying to lose weight, gluten can prevent your body from doing so. Overweight and obesity, after all, are also rooted in inflammation. And it's a two-way street: Inflammation promotes weight gain, and weight gain promotes inflammation. First, elevated inflammatory cytokines in the bloodstream, the hallmarks of inflammation, cause insulin resistance. This explains why people with other inflammatory conditions are at a higher risk of developing type 2 diabetes. Second, inflammation rages in the fat cells themselves upon the development of obesity. To be sure, body fat does serve a purpose; nobody can be totally fat-free. Fat alone is not an inflammatory tissue, but copious amounts of it beyond what's healthy for the body are problematic and trigger a self-perpetuating cycle of inflammation. This intracellular inflammation in fat tissue further promotes insulin resistance and weight gain.

Inflammation that takes place in the brain and gut exacerbates matters. Recall that leptin controls appetite and metabolism. When inflammation reaches the brain, specifically the hypothalamus, leptin resistance results, which then impairs glucose and fat metabolism. A similar scenario can happen in the gut: Inflammation of the gut causes leptin and insulin resistance, largely due to the exposure of toxins from the gut leaking into the bloodstream. One toxin in particular, lipopolysaccharide (LPS), is produced in the gut, where it belongs, by certain bacteria. Once LPS sneaks through the gut lining, however, it causes not only inflammation, but also insulin resistance in the liver and weight gain.

There are other connections between inflammation and

overweight/obesity. But the point I want to make is that gluten leads to a leaky gut, which then opens the door to the chronic inflammation that makes weight loss virtually impossible. I can't tell you how many people who have ditched gluten share stories of significant weight loss. As I've mentioned, this was not something I expected when I wrote my previous books.

In 2015 and 2016, newer research surfaced about the damaging effects of gluten on the microbiome as well. Indeed, it's quite possible that the entire cascade of adverse effects that takes place when the body is exposed to gluten starts with a change in the microbiome—ground zero. It goes without saying: You're going to evict this ingredient from your life. I'll show you how to do that in Part II.

GO LOW-CARB, HIGHER FAT AND FIBER

What's better for you, a low-carb or low-fat diet? Let's turn to the best medical literature available. In a Tulane University study, published in 2014 in the prestigious *Annals of Internal Medicine*, of 148 obese men and women who did not have cardiovascular disease or diabetes, half were placed on a low-fat diet and the other half on a low-carb diet. They were then followed for one year. The results were compelling: "The low-carbohydrate diet was more effective for weight loss and cardiovascular risk factor reduction than the low-fat diet. Restricting carbohydrate may be an option for persons seeking to lose weight and reduce cardiovascular risk factors." The people on the low-carb diet lost more weight, shrunk their waistlines to a greater extent, improved their cholesterol profiles (more good cholesterol, less bad), and enjoyed a dramatic drop in triglycerides, which are a strong risk factor for cardiovascular disease (CVD).

Now, why in a book about brain health would I address heart health? For starters, more than one-third of American adults have at least one form of CVD, and one-third of total deaths are due to CVD. The annual cost of caring for Americans with CVD is in the hundreds of billions and is projected to increase to approximately $1.48 trillion by 2030. Cardiovascular disease is one of the most important public health challenges in the United States. Second, both CVD and obesity are well-documented risk factors for brain disease. The common denominator is, of course, inflammation. In fact, in the low-carb vs. low-fat study, those on the low-carb diet experienced a drop in their levels of C-reactive protein, a blood marker for inflammation. Those on the low-fat diet, however, showed a *hike* in their C-reactive protein levels.

To me, these facts are staggering. Over the past sixty-odd years, we've learned over and over again that fat is fattening, and that avoiding traditional fats (such as olive oil, coconut oil, animal fats, nuts, avocados, and eggs) in favor of processed, manufactured fat substitutes is better for us and our waistlines. This led people to turn to a high-carb diet filled with sugars and synthetic fats, and the results have been disastrous.

The publication of a flawed study decades ago ignited the campaign against fat. In the 1950s, the University of Minnesota's Dr. Ancel Keys was determined to prove a correlation between the consumption of certain fats, particularly saturated fat and cholesterol, and cardiovascular disease. He charted the incidence of heart disease in various countries. Seeking to find a linear relationship, he removed some of the data points on his graph until he could see a clear pattern between fat consumption and heart disease. He left out countries that showed a paradox—countries like Holland and Norway where people eat a lot of fat but have little heart disease, and places like Chile where heart disease rates are high despite low-fat diets. What became known as the Seven Countries Study was not one that employed the rigors of the scientific method. But his contrived ideas stuck, and cholesterol became the villain.

Let me say a few quick things about this so-called villain. Cholesterol serves a critical role as a brain nutrient essential to the function of neurons. It is also crucial in building cellular membranes. Moreover, cholesterol acts as an antioxidant and a precursor to important brain-supporting molecules like vitamin D, as well as the steroid-related hormones (for example, sex hormones such as testosterone and estrogen). The brain demands high amounts of cholesterol as a source of fuel. All of the latest science shows that when cholesterol levels are low, the brain simply doesn't work well. People with low cholesterol are at much greater risk for neurological problems from depression to dementia.

But the food industry would have you thinking otherwise. When cholesterol became the bad guy, corporate food executives got to work to make and distribute hydrogenated butter-like substances, processed vegetable oils, and food products made with these horrible ingredients. They started labeling these products, which are teeming with dangerous trans fats, as "low in cholesterol" or "cholesterol-free." In the aftermath of this move from real food to manufactured food, we have suffered rising rates of chronic diseases rooted in inflammation, many of which are the very maladies we were hoping to prevent, such as diabetes and heart disease.

The idea that we should restrict our saturated fat is not supported by the latest dietary guidelines. In fact, when the new 2015 federal guidelines were published, most people—health "experts" included—were shocked to see that they had removed recommendations to limit the consumption of cholesterol-rich foods and added a nod to coffee as potentially being part of a healthy diet. Imagine that! The greatest risk to our health and to weight gain comes from replacing saturated fat with pro-inflammatory carbohydrates and sugars. We've got to welcome saturated fat back to the table. We've also got to embrace more natural fats in general, and not be afraid of a fat-driven diet. And we have to simultaneously lower our carb intake. High-fat and high-carb diets filled with gluten are the worst; not only do they wreak havoc on the metabolism and drive inflammation, but they also do a number on our gut bacteria. Study after study shows that the only way the high-fat diet works is when it is accompanied by low carbs; and the more fiber, the better. Remember, it's the fiber that feeds those gut bugs and contributes to intestinal health.

In Part II, I ask you to add more olive oil to your diet, especially in light of results from the recent PREDIMED (Prevención con Dieta Mediterránean) studies. They were conducted in Spain, and published in the American Medical Association's journal in 2015, to evaluate the effect of the Mediterranean diet vs. the effect of the low-fat diet recommended for people with breast cancer. Incidence of breast cancer has increased by more than 20 percent worldwide since 2008.

The Mediterranean diet is nutrient dense and low in sugars, and it welcomes an abundance of fat to the table. The studies, on more than 4,200 women aged sixty to eighty years, covered a six-year period. The women were split into three different groups: One group was placed on a Mediterranean diet with added mixed nuts. A second group was also on a Mediterranean diet, but with added olive oil. And the third group followed a low-fat diet. After 4.8 years, there were a total of thirty-five confirmed cases of breast cancer across all three groups. Breast cancer risk in the Mediterranean diet group with the added mixed nuts was 34 percent lower compared to risk in the low-fat diet group, while the risk for breast cancer in the Mediterranean diet group with added olive oil was an incredible 55 percent lower compared to the low-fat diet group.

If, through diet, you can guard against a disease as grave as cancer, a disease rooted in inflammation, imagine what else you can guard against. Other studies have echoed the PREDIMED project and arrived at the same conclusion. One in particular, published in the same journal in 2015, found that "a Mediterranean diet supplemented with olive oil or nuts is associated with improved cognitive function."

It's clear to me that the U-turn in our dietary choices over the past century is at fault for many of our modern scourges. As we went from eating a high-fat, high-fiber, low-carb diet to a low-fat, low-fiber, high-carb one, we concomitantly began to suffer from chronic conditions, many of which affect the brain. So get ready to eat like a hunter-gatherer. You will no longer fear dietary fat, even the saturated kind that's high in cholesterol. You will cut the carbs, and amp up the fat and fiber.

ABANDON SUGAR (REAL, PROCESSED, AND ARTIFICIAL)

Sugar is in almost every packaged food. It may be labeled differently—cane sugar, barley malt, crystalline fructose, evaporated cane juice, caramel, high-fructose corn syrup, maltodextrin—but it's all sugar (there are more than sixty names for sugar). Americans consume 22 teaspoons of sugar per day and more than 130 pounds (59kg) of sugar each year. In the past hundred years, there's been a fivefold increase in the consumption of fructose, found naturally in fruits but mostly consumed through highly processed foods containing high-fructose corn syrup. Fructose is implicated in the development of nonalcoholic fatty liver disease, a condition in which fat accumulates in the liver and triggers inflammation; it can also lead to scarring and cirrhosis. Consuming fructose, in fact, is associated with insulin resistance, high blood fats, and high blood pressure (hypertension). It is seven times more likely than glucose to result in sticky, caramel-like protein/carbohydrate aggregates called glycation end products, which cause oxidative stress and inflammation. Fructose does not prompt the production of insulin and leptin, the two key hormones that regulate metabolism, which is partly why diets high in fructose can lead to obesity and its metabolic consequences that indeed reach the brain and cause dysfunction. In fact, sugar causes adverse changes in our cellular membranes, our arteries, our hormones, our immune systems, our intestines, and our entire neurological system.

Oust the OJ: Would you drink a can of fizzy drink for breakfast? Probably not (though some do). When I ask audiences this question, I follow it with: What's the better choice, OJ or a can of regular Coke or Pepsi? They assume the former, but the truth is a 350ml glass of OJ contains 36 grams of carbohydrate, or 9 teaspoons of pure sugar, about the same as in a can of regular cola. But what about the vitamin C? Sorry, folks. The vitamin C in no way offsets the damaging effects of all that sugar. And if you think homemade is better, know that juicing in general is a bad idea. When fruits and vegetables are in their whole state, with all their fiber, the sugar is released slowly into your bloodstream, so your insulin response is tempered. In the juicing process, the fiber—the pulp—is strained out.

While we like to think we're doing ourselves a favor by replacing refined sugar with seminatural products like Truvia and Splenda (which are marketed as "made from nature"), these are processed chemicals in disguise. What about artificial sweeteners? The human body cannot digest these, which is why they have no calories. But they must still pass through the gastrointestinal tract. For a long time, we assumed that artificial sweeteners were, for the most part, inert ingredients that didn't affect our physiology. Far from it. In 2014 a watershed paper, which has since been widely referenced, was published in *Nature* proving that artificial sweeteners affect gut bacteria in ways that lead to metabolic dysfunction, such as insulin resistance and diabetes, contributing to the same overweight and obesity epidemic they were marketed to provide a solution for.

Be On the Lookout: Examples of Popular Sugars and Sweeteners

evaporated cane juice

corn syrup

high-fructose corn syrup

crystalline fructose

fructose

sucrose

malt

maltose

maltodextrin

dextrose

beet sugar

turbinado sugar

invert sugar

aspartame

cyclamate

saccharin

sucralose

AVOID GMO FOODS

A lot of research is currently under way to study the effects of genetically modified organisms (GMO) on our health and on the environment. GMOs are plants or animals that have been genetically engineered with DNA from other living things, including bacteria, viruses, plants, and animals. The genetic combinations that result do not happen naturally in the wild or in traditional crossbreeding. GMO foods are commonly created to fight louses and viruses that can destroy crops, or to cultivate crops with certain desired characteristics. In the 1990s, for example, the ringspot virus decimated nearly half the crop of Hawaiian papaya in the state. In 1998, scientists developed a genetically engineered version of the papaya called the Rainbow papaya, which is resistant to the virus. Now more than 70 percent of the papayas grown in Hawaii are GMO.

> Corn and soy are the top two GMO crops in the United States, and it's estimated that GMOs are in as much as 80 percent of conventional processed foods. Restrictions or complete bans have been placed on the production and sale of GMOs in more than sixty countries worldwide, including Australia, Japan, and all of the countries in the European Union. But in the United States, the government approves them. The problem: Many of the studies showing GMOs to be safe have been performed by the same corporations that created and now profit from them. People are rallying around the country for better food labeling so they can choose to opt out of what some call "the experiment."

There are a variety of genetically modified or engineered foods that the GMO industry features when trying to convince us of the merits of this technology. There is a sweet potato grown across Africa that has been engineered to be resistant to a particular virus. Rice has been engineered to increase its vitamin and iron content. Plants are genetically modified to be more resistant to various extremes of weather. There are fruit and nut trees that are engineered to yield crops years earlier than they would normally. Bananas are even genetically modified to produce human vaccines against diseases like hepatitis B. This all sounds promising, especially in light of issues of food scarcity in developing countries. But the story is not yet complete. While it's true that not all genetically modified organisms are inherently bad, the methods used to create and farm GMOs can entail practices with far-reaching consequences, many of which we don't understand yet.

We've been told, for example, that the new AquAdvantage salmon (made by AquaBounty Technologies) is safe for human consumption. But the FDA has only looked at the effect of this genetically modified fish on the environment. No studies have been performed to examine its effects on humans. We know that gene modification changes specific proteins, and that proteins we consume affect our own gene expression. You will not find any study on how the consumption of this fish changes gene expression in humans who eat them. According to Jaydee Hanson, senior policy analyst at the Center for Food Safety, "The modus operandi at FDA is to rubber stamp AquaBounty's studies and then call its review process science-based." Hanson goes on to say, "FDA's inadequate risk assessment is at odds with reality, with science and with the public, which has long called on the agency to put consumers' health and environmental safety ahead of the corporate interests of the biotechnology industry."

In a scathing review of GMOs by Consumer Reports, Robert Gould, MD, president of the board of Physicians for Social Responsibility, was quoted as stating: "The contention that GMOs pose no risks to human health can't be supported by studies that have measured a time frame that is too short to determine the effects of exposure over a lifetime." He then goes on to call for more studies to assess GMOs' long-term effects, especially given the fact that animal studies show they might cause damage to the immune system, liver, and kidneys. He also points out that a lack of labeling prevents researchers from even tracking the potential health effects of GMOs.

In addition to concerns about the effects of altered genetics in GMOs on human health, one of the most problematic—and contentious—aspects of GMOs has to do with current farming practices to grow GMO foods. No longer do farmers yank weeds from their fields by hand or machinery. They now spray a weed-killing chemical, glyphosate (the active ingredient in the common herbicide Roundup), on their crops. And they apply even more of this chemical just before the harvest to obtain a bigger yield and as a drying agent to prime the soil for a new crop. U.S. farm workers have sprayed 1.8 million tons of glyphosate since Roundup came on the market in 1974. Globally, 9.4 million tons of the chemical have been sprayed onto fields. It's estimated that by 2017 American farmers will apply an astounding 1.35 million metric tons of glyphosate to their crops. That's just under 3 billion pounds.

In order to protect crops from the herbicide, the seeds are genetically modified to be resistant to the herbicide's effects. In the world of agriculture, these seeds are known as "Roundup-ready". The use of Roundup-ready GMO seeds has allowed farmers to use huge amounts of this herbicide. Which means GMO foods—and foods conventionally farmed—are invariably contaminated with glyphosate, the "tobacco" of the 21st century that wreaks havoc on human health. Farmers who grow organic foods fear contamination in their fields as well. Glyphosate is a poison like no other, toxic to the gut all the way up to the brain.

Many of glyphosate's adverse effects are found at very low doses, challenging the notion that there is such a thing as a safe threshold of exposure. An entire book could be written about the politics and the biological effects of glyphosate. But for now, let me point out the main concerns as they relate to human health.

Glyphosate:

- acts as a powerful antibiotic, slaughtering beneficial bacteria in your gut and thereby disrupting the healthy balance of your microbiome
- mimics hormones like oestrogen, driving or stimulating the formation of hormone-sensitive cancers
- impairs the function of vitamin D, an important player in human physiology
- depletes key compounds like iron, cobalt, molybdenum, and copper
- compromises your ability to detoxify toxins
- impairs the synthesis of tryptophan and tyrosine, important amino acids in protein and neurotransmitter production

It wouldn't surprise me in the least if it were soon revealed that the obesity epidemic could partly be blamed on the widespread use of glyphosate and consumption of GMOs due to the chemical's effects on gut health and the microbiome. The importance of avoiding foods that have come into contact with glyphosate cannot be overemphasized. It can be found in unlikely places. In 2015, for instance, it was detected in PediaSure Enteral formula, which is widely used by hospitals in the United States for children in intensive care in need of nutrition. It is used in the wine industry. It has even been found in sanitary products because it is used by the cotton industry.

We must stand up in protest against this unacceptable experiment. Until glyphosate is banned, you'll need to focus on organic produce, pastured animal foods, and non-GMO verified products.

There is now a urinary test to measure glyphosate. This is one of the tests you might consider taking (see page 72 for more).

WATCH OUT FOR TOO MUCH PROTEIN

Picture yourself at a dinner party with friends. That morning, a headline in the media reported on the health risks surrounding red meat. This particular report spread like wildfire in the media because it came from Harvard's School of Public Health. According to the study, for every extra serving of unprocessed red meat consumed daily beyond an acceptable single serving (about the size of a deck of cards), the risk of dying prematurely went up by 13 percent; one daily serving of processed red meat, such as one hot dog, two slices of bacon, or one slice of cold cuts, increased the risk by 20 percent. You love your steak, but a dinner mate across the table—a vegetarian—decides to debate you and it gets a little heated. Who is right?

I am frequently asked about meat consumption. The Harvard study I just referenced was no small study, and to date it has been the largest, longest study on the alleged link between red meat and life span. It included data from two studies of more than 37,000 men and 83,600 women. These volunteers were followed for an average of twenty-four years, during which a combined 23,926 of them died—5,910 from cardiovascular disease and 9,464 from cancer. Every four years, they submitted information about their diets. Generally speaking, those who ate the most red meat had higher death rates compared to those who ate the least. In particular, in those who ate one serving of red meat per day or more, the corresponding increases in risk for both men and women were 18 percent and 21 percent for cardiovascular death, and 10 percent and 16 percent for cancer death. These analyses factored chronic disease risk factors into the equation, including age, family medical history for heart disease and cancer, body mass index, and amount of physical activity. While the findings have merit, they do not tell the whole story. These are statistical links, after all.

If eating red meat hiked your risk of dying prematurely by as much as 21 percent, that might encourage you to learn to love tofu and tempeh instead of steak and bacon. But we're talking about the *relative* risk of eating more red meat versus less. It behooves us to consider the *absolute* risk, which brings those percentages down considerably—to single digits. Complicating this already complex story is the fact that big red meat eaters often have other risk factors for serious, life-shortening diseases. While seemingly stereotypical, data shows that people who eat too much red meat also tend to avoid exercise, drink alcohol excessively, and smoke. The researchers tried to compensate for the effects of unhealthy lifestyles in their study, finding that mortality and meat consumption remained associated, but with so many variables, it's hard to tease out hard numbers that are meaningful. And it's harder still to apply those statistics to everyone. The effects of unhealthy choices are relative to where you start—how old you are, how long you've been engaging in certain habits, and what your underlying risk factors are from a genetic standpoint. This muddies the waters even further.

An important conclusion in the study was that the higher death rate could have been mitigated had meat consumption been limited to less than half a serving per day. That's less than three-and-a-half servings per week. Moderation wins. Meat is not necessarily bad, particularly when it's not processed. And therein lies the key to enjoying red meat: Choose high-quality meat that has not been treated with antibiotics or force-fed genetically modified grains sprayed with glyphosate. I bet if a study were done comparing people who eat conventional meat to those who eat grass-fed organic meat, you'd see a difference in their health risks, despite how much meat they are consuming! High-quality meats come with high-quality fats, too.

Let me touch upon that other misconception I often hear. Contrary to what you might think, going low-carb does not mean going high-protein. You will not be eating meat every day. A lot of people think they need upward of 100 grams of protein a day, but in fact we only need about half that (see next page). Vegetarians often ask me if they should worry about getting enough protein, and I reassure them that they consume plenty from plant sources, legumes, eggs, nuts, and seeds. You probably eat more than enough protein each day, but you could be consuming too much. Protein is an essential component of any diet, but more doesn't mean better or healthier. Extra protein will not help you burn more fat, build more muscle, or render you stronger. If you consume too much of it, you'll eat more calories than your body requires, store more fat, and put yourself at risk for an earlier death.

A 2014 study conducted at multiple centers around the world demonstrated the value in reducing protein consumption for longevity. Those individuals with the highest protein intake during the eighteen-year study had a fourfold increased risk of dying from cancer and a fivefold increased risk of dying from diabetes (note that some of this increased risk could be attributed to too much animal protein in particular). But here's what they also found: The increased risk of cancer death was identified in people aged fifty to sixty-five years old who were consuming a high amount of protein. In people over the age of sixty-five, the trend reversed: These people had a *reduced* cancer risk (but still the same fivefold increased risk of diabetes mortality). So what do we make of this? The study concluded: "These results suggest that low protein intake during middle age followed by moderate to high protein consumption in old adults may optimize health span and longevity."

According to the Centers for Disease Control and Prevention, we only need to get 10 to 35 percent of our day's calories from protein foods; this translates to about 46 grams of protein for women and 56 grams of protein for men. It's easy to get your daily requirement when you consider the following: A 75g piece of meat has about 21 grams of protein (if you eat a 225g piece of meat, it could contain more than 50 grams).

More Isn't Better: You Need Less Protein Than You Realize

Low-carb does not mean high-protein. The Grain Brain Whole Life Plan calls for a limited daily intake of protein—no more than about 46 grams for women and 56 grams for men.

On the Grain Brain Whole Life Plan, you're going to rejoice in the delicious proteins you get to enjoy, and you won't feel deprived. To ensure you get just the right blend of different proteins and their amino acid makeup, you'll be mixing up the types of protein you eat. And one source of high-quality protein that you'll love as much as I do is the incredible egg.

EMBRACE THE INCREDIBLE EGG

Eggs are a staple in my diet. I panic when I run out of eggs. I don't panic at the thought that they are a high-cholesterol food. Remember, the idea that dietary cholesterol, such as that in saturated fat from beef, converts directly into blood cholesterol is totally false. Science has never been able to connect dietary animal fats and dietary cholesterol to levels of blood cholesterol or risk of coronary heart disease. And when scientists try to track a relationship between blood cholesterol and egg consumption, they continually document that cholesterol levels in people who eat few or no eggs are identical to people who consume lots of eggs. More than 80 percent of the cholesterol in your blood is produced by your own liver, and, contrary to what you might think, consuming cholesterol has been shown to actually *reduce* your body's production of cholesterol.

Eggs—yolks included—are an unrivaled food. They are versatile, and cost-effective, nutritional gold mines. I invite you to check out my online video about eggs when you have a moment at www.DrPerlmutter.com. Whole eggs contain all of the essential amino acids we need to survive, vitamins and minerals, and antioxidants known to protect our eyes. And they can have far-reaching positive effects on our physiology. Not only do they keep us feeling full and satisfied, but they also help us control blood sugar and, in turn, a whole panoply of risk factors for illnesses as diverse as heart disease, cancer, and brain-related disorders.

You'll see that I recommend lots of eggs in the plan. I find they are a perfect way to start the day; they give you the ideal combination of fat and protein to set your body's biological "tone" for the day. Please don't be afraid of them, especially the "high-cholesterol" yolks. But, as with other sources of proteins, choose your eggs carefully. Pastured eggs, which come from chickens that are allowed to roam free and eat what they normally would in the wild (plants and insects, not processed grains), are best. They taste better, too! There are so many things you can do with eggs. Whether you scramble, fry, poach, boil, or use them in dishes, eggs are indeed among the most multipurpose ingredients.

IT'S MORE THAN JUST FOOD

I highlighted a lot of "rules" in this chapter that cover nutrition. But as you'll see coming up, it goes far beyond that. Nobody changes his or her health for the better solely through food. Although implementing the ideas in this section will set you free in a lot of ways, there's much more to consider in terms of what factors into your health and risk for brain disease: how you reduce the stress in your life, how well you sleep at night, the drugs you reach for because you think you need them, and whether or not you think highly of yourself and the people around you. I'll be giving you the practical tools and strategies to address every angle of your lifestyle. Now let's get to Part II.

PART II

THE GRAIN BRAIN WHOLE LIFE PLAN ESSENTIALS

It's been fifteen months since I made a forever change to my diet and lifestyle by converting to a low-carb, gluten-free, high-fat diet. From a starting point of a fleshy 225 pounds, I am now down to 198 pounds and loving it! I started small, by eliminating all sugars, carbohydrates, and gluten for thirty days. I lost 3 pounds per week, even with no change to my gym schedule, which was sporadic at best. My jeans that were once tight were falling off my butt and I had to wash them in hot water and shrink them. For the first time in a long time, I was happy to look at myself in the mirror after a shower—heck, at forty-eight years young, I look better than I did when I went to the gym two hours a day when I was twenty!

—Pat L.

CHAPTER 4

Getting Started: Assess Your Risk Factors, Know Your Numbers, and Prepare Your Mind

TIME TO TURN THE SCIENCE into success. I've given you a great deal of information up to this point. You've learned a lot about the biology of health in the 21st century, some of which may have turned what you believed as conventional wisdom on its head. If you haven't already begun to change a few things in your life based on what you've read, now is your chance. In Part II, you'll learn how to shift your lifestyle and bring your body—and brain—back to optimal well-being. You'll feel energetic and vibrant, suffering less from chronic conditions.

Making lifestyle changes, even small ones, can seem overwhelming at first. How will you be able to resist your usual habits? Will you feel deprived and hungry? Will you find it impossible to keep up this new lifestyle forever? Is this program doable, given the time and commitments you have? Can you reach the point where following these guidelines is automatic, like second nature?

Take a deep breath. You will soon have even more knowledge and inspiration to help you stay on a healthy path for the rest of your life.

The closer you stick to my guidelines, the faster you will see results (and maintain them!). Bear in mind that this program has many benefits beyond the obvious physical ones. Ending fears of cognitive decline might be first and foremost on your mind, but the rewards don't stop there. You will see change in every area of your life. You will feel more confident and have more self-esteem. You'll feel younger and more in control of your life and future. You'll be able to navigate through stressful times with ease, have the motivation to stay active and engage with others, and feel more accomplished at work and at home. In short, you will feel and *be* more productive and fulfilled.

And your success will propagate more success. When your life becomes better, fuller, and more energized, you won't want to go back to your old, unhealthy lifestyle. I know you can do this. You must, for yourself and your loved ones. The payoffs are huge.

Let's get started with a quick rundown of the entire program:

Prelude: *Assess* Your Risk Factors, Know Your Numbers, and Prepare Your Mind

- Assess your risk factors using the quiz on page 70
- Have lab tests performed using the guide on page 72
- Turn off your autopilot (see page 76) and consider fasting for a day

Step 1: *Edit* Your Diet and Pill-Popping

- Learn how to nix the villains in your diet (see page 88) and welcome in the heroes (see page 93) that will help support the structure and function of your body

- Know which supplements you should consider adding to your daily regimen (see page 102) and the medications you should try to dump if possible (see page 113)

Step 2: *Add* Your Support Strategies

- Establish an exercise routine you can sustain (see page 120)
- Pay attention to pain, especially in your back and knees (see page 132)
- Make room for sleep (see page 136)
- Reduce stress and find calm in four simple ways (see page 142)
- Detoxify your physical environment (see page 159)

Step 3: *Plan* Accordingly

- Know when to eat (see page 166), sleep (see page 170), and exercise (see page 168). Train yourself to plan your days so that you achieve your daily goals given your time constraints and responsibilities. Be ruthless with your schedule and your realities

Now, on to the prelude, which will gear you up for Step 1

ASSESS YOUR RISK FACTORS

The quiz below will arm you with some personal data that can help provide a sense of your risk factors for brain disorders and disease, which can manifest in migraines, seizures, mood and movement disorders, sexual dysfunction, and ADHD, as well as serious mental decline in the future.

Remember, the organs and systems of the body are highly interconnected and intertwined. If this quiz determines that you're at higher risk for brain disease, then it also means you're at greater risk for a medley of other illnesses that are not, in and of themselves, brain related.

Respond to these statements as honestly as possible (*Y* means yes; *N* means no). If you don't know the answer to a question, skip over it.

1. Do you suffer from depression or chronic anxiety? Y/N
2. Were you born via C-section? Y/N
3. Are you more than twenty pounds (9kg) overweight? Y/N
4. Have you taken antibiotics at least once in the past year? Y/N
5. Do you avoid exercise? Y/N
6. Do you consume artificial sweeteners at least once a week (found in diet fizzy drinks, sugar-free gum, and other foods or products labeled "sugar-free")? Y/N
7. Are you on a low-fat diet? Y/N
8. Have you been diagnosed with a sleep disorder or do you suffer from insomnia? Y/N
9. Do you take proton pump inhibitors (Prilosec, Nexium, Prevacid) once in a while for heartburn or acid reflux? Y/N

10. Do you eat GMO foods such as nonorganic corn and soy? Y/N
11. Do you feel like you don't handle stress well? Y/N
12. Do you have a blood relative who has been diagnosed with Alzheimer's disease or coronary artery disease? Y/N
13. Is your fasting blood sugar 100 mg/dL or greater? Y/N
14. Have you been diagnosed with an autoimmune disorder (for example, Hashimoto's thyroiditis, Crohn's disease, rheumatoid arthritis, lupus, inflammatory bowel disease, multiple sclerosis, type 1 diabetes, psoriasis, Graves' disease)? Y/N
15. Do you sometimes take laxatives? Y/N
16. Do you take a nonsteroidal anti-inflammatory (for example, ibuprofen, naproxen) at least once a week? Y/N
17. Do you have type 2 diabetes? Y/N
18. Are you extra sensitive to chemicals often found in everyday products? Y/N
19. Do you have food allergies or are you gluten sensitive? Y/N
20. Do you eat bread, pasta, and cereal? Y/N

Don't be alarmed if you find yourself answering "yes" to most of these questions. The more yeses you have, the higher your risk for having dysfunctional physiology that might be impacting your health. But you are not doomed. The whole point of this book is to empower you to take charge of your health like never before.

KNOW YOUR NUMBERS: BASELINE BLOOD WORK

I recommend that you schedule the following lab tests, or as many of them as you can, as soon as possible. You can certainly start this program today while you wait for your appointment with a clinician and get the results, but knowing your numbers will not only motivate you to move forward but will also help you set target goals for each test result. You'll know where your biological weaknesses are, so you'll be able to pay attention to improving those numbers.

I've included healthy target levels where appropriate. Note that some of these tests are not normally performed by traditional doctors, so you may need to seek additional help from a functional medicine practitioner to complete these tests (see my website, www.DrPerlmutter.com, for details).

Fasting insulin: If you get only one test on this list, make it this one. It is critically important, and any health-care provider can perform it. Long before blood sugar begins to climb as a person becomes diabetic, the fasting insulin level will rise, indicating that the pancreas is working overtime to deal with the excess of dietary carbohydrates. It is a very effective early warning system for getting ahead of the diabetes curve, and so has tremendous relevance for preventing brain disease. You want this number to be below 8 uIU/ml (ideally, below 3).

Fasting blood glucose: A commonly used diagnostic tool to check for prediabetes and diabetes, this test measures the amount of sugar (glucose) in your blood after you have not eaten for at least eight hours. A level between 70 and 100 mg/dL is considered normal, but don't be fooled. A blood sugar pushing 100 is anything but normal. At that level, you're showing signs of insulin resistance and diabetes, and you have a heightened risk for brain disease. Ideally, you want to have a fasting blood glucose of less than 95 mg/dL.

Hemoglobin A1C: Unlike the fasting blood glucose test, this test reveals an "average" blood sugar over a ninety-day period and provides a far better indication of overall blood sugar control. Specifically, it measures the amount of *glycation* that the protein hemoglobin has undergone. Glycation, as I defined earlier, simply means that sugar has become bonded to a protein, in this case hemoglobin. This is a relatively slow process, but glycated hemoglobin is a powerful predictor of risk for Alzheimer's disease, as well as being one of the greatest predictors of brain shrinkage. A good A1C value is between 4.8 and 5.4 percent. Note that it can take time to see this number improve, which is why it's typically only measured every three to four months.

Fructosamine: Similar to the hemoglobin A1C test, a fructosamine test is used to measure an average blood sugar level but over a shorter time period—two to three weeks. Your fructosamine level should be between 188 and 223 µmol/L. It's possible to see positive changes in this test within two to three weeks.

Glyphosate urine test: Glyphosate, you'll recall, is the active ingredient in the popular weed killer Roundup that's used extensively in conventional farming today. We finally have a way to test for this man-made chemical in the body through a new urine test. You want to test negative for detectable levels of glyphosate in your urine, measured as ug/L.

C-reactive protein: This is a marker of inflammation in the body. You want to see 0.00 to 3.0 mg/L (ideally, less than 1.0 mg/L). CRP may take several months to improve, but you may well see positive changes even after one month following my protocol.

Homocysteine: Higher levels of this amino acid, produced by the body, are associated with many conditions, including atherosclerosis (narrowing and hardening of the arteries), heart disease, stroke, and dementia. It is now generally regarded as being quite toxic to the brain. Having a homocysteine level of just 14—a value exceeded by many of my patients when first examined—was described in the *New England Journal of Medicine* as being associated with a *doubling* of the risk for Alzheimer's disease (an "elevated" homocysteine level is anything above 10 µmol/L in the blood). High levels of homocysteine have also been shown to triple the rate of telomere shortening (remember that telomeres are those caps on the ends of your chromosomes that protect your genes and whose lengths are a biological indication of how fast you are aging). Homocysteine levels are almost always easy to improve (see below). Your level should be 8µmol/L or less. Both vitamin D and omega-3 fats can lengthen telomeres by increasing the activity of telomerase, the enzyme involved in lengthening telomeres. Many drugs can inhibit the B vitamins and raise homocysteine (see the list at www.DrPerlmutter.com under Resources), but most people can immediately correct their level just by supplementing with some B vitamins and folic acid. Typically, I ask patients with a poor homocysteine test to take 50 milligrams of vitamin B_6, 800 micrograms of folic acid, and 500 micrograms of vitamin B_{12} daily and retest after about three months.

Vitamin D: This is now recognized as a critical brain hormone (remember, it's not actually a vitamin; see page 106 for more details). Interestingly, higher levels of vitamin D are also associated with longer telomeres, a good thing. Your vitamin D level will probably be low (normal is between 30 and 100 ng/mL, but you ideally want to be around 80 ng/mL). The majority of Americans are deficient in this critical nutrient due to indoor living and the use of sunscreen; those living in northern latitudes are most at risk for being deficient.

Because it can take time for the body to shore up its levels of vitamin D upon supplementation, you'll start with 5,000 international units (IU) of vitamin D once a day, and test your level after two months. If after two months your level is 50 nanograms per milliliter (ng/mL) or under, you'll take an additional 5,000 IU daily and retest in two months. It's the level maintained in your body that matters, not the dosage. Ask your health-care practitioner to help you adjust your dosage to achieve an optimal level. Once you do, a daily dose of 2,000 IU will usually suffice to maintain a healthy level, but ask your doctor for specific recommendations.

Once you've been following my protocol for a couple of months, it's a good idea to have these laboratory studies repeated to measure improvements. It can take time to see dramatic changes in some of these parameters, but if you follow the plan to a T, you should nonetheless begin to feel positive changes within a few weeks, and that will inspire you to keep going.

PREPARE YOUR MIND

I realize that some of you might be a little worried at this juncture. Given the self-assessments and lab work you've hopefully completed (or soon will), maybe you think that the cards are stacked against you. And the thought of cutting carbs is adding more unwanted stress. I'm here for you, which is why we're going to consider three things to prepare your mind to go forward with resolve and the understanding that you have control and can change those cards so they are stacked in your favor.

Turn Off Your Autopilot

Rituals. Traditions. Habits. Ruts. We all have them. Some of them are good and help us stay healthy and fit. But some of them keep us running in the other direction and getting stuck. Do you find yourself rising in the morning in a semi-fog, inhaling a carb-filled breakfast without much thought, rushing throughout your day drinking fizzy drinks and calorie-laden coffee drinks, coming home exhausted and wishing you had energy to exercise, eating dinner mindlessly in front of the television, and then falling into bed? Is your life automated so much that one day seamlessly and monotonously becomes another?

If so, don't feel bad. You're reading this book to get out of your rut—that comfort zone that in the long run won't be so comfortable. You don't want to look at yourself in the mirror five or ten years from now and be twenty pounds heavier, a hundred times more miserable, and on the road to experiencing serious health issues, if you don't have them already. My guess is you have your staples—your favorite foods, restaurants, routines, and shortcuts in every aspect of life. Now is the time to awaken to a new *whole life*.

It's important that we learn how to turn off our autopilot. I'm going to help you do that with the strategies in the rest of the book. The moment you begin to 1) edit your diet and pill-popping; 2) add your support strategies; and 3) plan accordingly, you'll start to avoid autopilot and turn instead to a much more fulfilling, energizing life. Turning off your autopilot will also happen automatically once you flip a few switches in your body by jump-starting the program with a fast and then cutting carbs cold turkey.

Fast and Furious (Should You Jump-Start with a Fast?)

If you've gone on diets before, then at one point you were probably told to eat five or six small, healthy meals during the day to keep your metabolism in gear. You were persuaded to believe that eating this way supported calorie burning, and that any sensation of hunger triggered alarms in the body to store fat and slow down metabolism.

We might have come a long way in our technological advancement, but from an evolutionary standpoint, our DNA isn't much different from the DNA of our hunter-gatherer ancestors. And contrary to what you may have been told, our ancestors didn't eat six times a day. For them, it was feast or famine. They needed to be able to endure long periods of time without food.

Plato was right when he said, "I fast for greater physical and mental efficiency." And so was Mark Twain when he declared, "A little starvation can really do more for the average sick man than can the best medicines and the best doctors." Many religions encourage the practice of fasting as a spiritual practice. There's the Islamic fast of Ramadan, the Jewish fast of Yom Kippur, and a variety of centuries-old fasting practices in Christianity, Buddhism, Hinduism, and Taoism. Although there are many different types of fasting, in general all fasts share one thing: They call for a willing abstinence from or a reduction of food, drink, or both for a period of time.

Fasting is a long-established way of physically rebooting the metabolism, promoting weight loss, and even increasing mental clarity and insight. (This latter fact makes sense from an evolutionary standpoint: When food was scarce, we needed to think quickly and smartly to find our next meal!) The scientific evidence for its benefits has been accumulating. In the early part of the twentieth century, doctors began recommending it to treat various disorders such as diabetes, obesity, and epilepsy. Today we have an impressive body of research to show that intermittent fasting, which includes everything from seasonal fasts lasting a few days to merely skipping a meal or two routinely on certain days of the week, can increase longevity and delay the onset of diseases that tend to cut life short, including dementia and cancer. And despite popular wisdom that says fasting slows down the metabolism and forces the body to hold onto fat in the face of what it perceives as starvation mode, it actually provides the body with benefits that can accelerate and enhance weight loss.

Typically, our daily food consumption supplies the brain with glucose for fuel. Between meals, the brain receives a steady stream of glucose made from glycogen, stored mostly in the liver and muscles. But glycogen reserves can provide only so much glucose. Once they are depleted, our metabolism shifts and we create new molecules of glucose from amino acids taken from the protein primarily found in muscle. On the plus side, we get more glucose; on the minus side, it comes at the expense of our muscles. And muscle breakdown is not a good thing.

Luckily, our physiology offers one more pathway to power the brain. When quick sources of energy like glucose are no longer available to fuel the body's energy needs, the liver begins to use body fat to create ketones, specialized molecules I described in Part I. One ketone in particular plays a starring role: beta-hydroxybutyrate (beta-HBA). Beta-HBA serves as an exceptional fuel source for the brain. This alternative fuel source allows us to function cognitively for extended periods during food scarcity. It helps reduce our dependence on gluconeogenesis and, therefore, preserves our muscle mass. Gluconeogenesis is the process by which the body creates new glucose by converting non-carbohydrate sources, such as amino acids from muscles, into glucose. If we can avoid the breakdown of muscle mass for fuel and instead utilize our fat stores with the help of ketones like beta-HBA, that's a good thing. And fasting is a way to achieve this goal.

Fasting also powers up the Nrf2 pathway I discussed in Chapter 2, leading to enhanced detoxification, reduction of inflammation, and increased production of brain-protecting antioxidants.

Despite all the benefits of fasting I just described, perhaps one of the best outcomes of the practice, especially during the prelude period of this program, is that it can help you mentally prepare for the dietary protocol. If you're worried about what it will be like to drastically reduce your carbohydrate consumption overnight, then I can't think of a better way to equip your mind—and body—for that achievement than to fast for a twenty-four-hour period before commencing the program. I also recommend that regardless of your medical status and history, you check with your doctor first before fasting for any length of time. If you take any medication, for example, ask about continuing to do so during your fast.

So unless you have a medical condition that prevents you from fasting, make it a goal to try the following: .

Fast for one full day: Before you commence my 14-day meal plan (see page 200), set the foundation mentally and physically by drinking filtered water only for the twenty-four-hour period leading up to that first meal. For many, it helps to do the fast on a Saturday (last meal is dinner Friday night), and then begin the diet program on a Sunday morning. A twenty-four-hour fast is also a great way to jump back into this way of life if you fall off the proverbial wagon.

Skip breakfast once in a while: The body wakes up in mild ketosis. If you skip breakfast, you can keep this state going for a few hours before eating lunch at midday. Try skipping breakfast once or twice a week. I'll be asking you to do that during the 14-day meal plan. If the days I ask you to skip breakfast are not ideal for you, then choose one or two other days during the week when it works for you.

Fast for a full seventy-two hours: Four times a year, go on a prolonged fast of seventy-two hours, during which you only drink filtered water. This type of fast is more intense, as you can imagine, so be sure that you've tried a few twenty-four-hour fasts before attempting this one. Fasting during the seasonal changes (the last week of September, December, March, and June) is an excellent practice to maintain.

Go Cold Turkey (Cutting Off the Carbs)

Okay, you're almost ready for Step 1. I know what you're thinking. The thought of going cold turkey on cutting carbs is terrifying. Let me share Jen Z.'s story and then offer some advice.

My name is Jen, I am fifty-four years old, and for quite a number of years I had been dealing with some very questionable illnesses. I was overweight with no success in getting my weight down, was chronically tired, had trouble concentrating, and had developed an autoimmune skin disorder called vitiligo, where the skin no longer produces pigment. A couple of years into these problems, I was diagnosed with metastatic melanoma and went through surgery and very aggressive immunotherapy as well as radiation. The treatment for the cancer left me with very damaged nerves, damaged skin, no energy at all, arthritis and joint pains so bad I could hardly walk sometimes, and the brain power of a pea. I could not remember things that I had known all my life, nor could I concentrate on anything.

With all my research trying to figure out how long this would last, I was led to the conclusion that this would be my new normal and I would just have to deal with it. I am an avid horsewoman, and this left me feeling that everything I had worked toward was being taken away.

Then one day about seven months ago, a friend shared with me an article about a doctor who was diagnosed with untreatable stage 4 brain cancer. In his quest to fight for his life, he discovered the low-carb lifestyle. I say lifestyle because it is not a diet to be jumped on and off of but a new way of life. At this point, I was searching for a way to regain some of my health and try to avoid any need for cancer treatments, so I thought I would give it a try. The cancer version of this new lifestyle is quite extreme, but it has been so worth it! The first two weeks were pretty rough with food cravings and my body adjusting to the change in diet, but I could feel changes happening and they felt good. I went cold turkey, all in! About four weeks in, my body no longer hurt from arthritis pain, joint swelling was gone, and my weight was dropping without me doing anything but changing the way I eat. Jumping ahead seven months now, I have been very faithful to the program! Only a handful of times have I put food in my body that does not belong and it was not but a few bites. Does not take much and you know that it was not a smart choice!

Anyway, I feel better now than I have in probably twenty years or longer. My brain function is back better than before, nerves are regenerating that were destroyed by radiation, the skin disorder that I was told would get progressively worse is actually reversing, my energy level is crazy good, and I am back training horses again. Best of all, I am three years cancer-free, my weight is almost what it should be, and I feel great. I will be the first to say it was not simple getting started, but once you get it figured out (there are lots of food labels to read), it gets easier by the day.

The vast majority of you will have no trouble going cold turkey on cutting carbs. But it may be difficult for some, especially if carbs have been a large part of your diet. If you experience mood swings, crashing energy levels, and intense cravings during the first couple of days on the program, have patience. These effects are temporary and will go away within the first week. Your mind will clear, your energy levels will soar, and you'll realize how important this decision has been.

You'll never want to go back. Here are some additional thoughts to consider while facing the task of nixing those addictive carbs.

Leverage your motivation: Sugar and drugs have a lot in common. Cravings for both act on the same neurochemical pathways, which is why weaning yourself off either drugs or sugar in the form of processed carbs can involve unwanted withdrawal effects (though withdrawing from sugar is easier than withdrawing from most drugs). As I just mentioned, many people don't have a hard time cutting carbohydrates, but you might experience a short-lived "thirst" for sugar, crankiness, headaches, low energy, maybe some aches and pains. This is normal. When you understand that the discomfort is merely a side effect of your body's natural withdrawal from an addictive substance, you can use the knowledge to allay your fears and frustrations, turning that knowledge into a source of motivation and resolve. Remind yourself that these effects are temporary and won't last long. When you don't feel great and have the urge to eat a carbohydrate-rich food that is calling to you like a best friend, talk yourself out of it. You will not let carbs control you like a drug. Think about how much better you'll feel when you expel them from your diet.

Arm yourself with alternatives: Let's be honest. Going from a carb-rich diet to a carb-less diet overnight can be a significant lifestyle change. Recognize that. Acknowledge that this way of eating can take a little while to get used to, and that's okay. During the first few days of your transition, arm yourself with the supplies for a counterattack when cravings hit by always having high-quality snacks on hand, such as raw nuts and nut butters, beef jerky, tasty cheeses, hard-boiled eggs, and raw veggies with a delicious dip (see page 198 for more snack ideas). Don't worry about counting calories or eating too many snacks. Just do it to get through the transition and I promise that you will emerge healthier, happier, and a whole lot lighter—and the cravings will soon vanish.

Avoid temptations: Say goodbye to some of your favorite restaurants. The hardest part of this carb-conscious transition is the beginning. Don't sabotage yourself or make it any harder by patronizing restaurants and food courts where you know that you'll be tempted and will have a difficult time finding something to eat that satisfies this lifestyle. Set yourself up for success at the start by avoiding unnecessary temptations. Of course, all within reason. You have commitments to fulfill and places to go, including events at your children's school, your work, and occasions in your personal, social world. Life does go on around you, so be mindful of that. Engineer the start of your new way of living when you know you can make a real go of this. If you have a work-related breakfast function on Friday morning that you know will involve a buffet filled with pancakes, doughnuts, and waffles, get past that and plan to start the protocol on Saturday. During the 14-day meal plan, bring lunch to work so you're in control and you're not stuck deciding between unhealthy options.

Take the challenge: Make the commitment today that you'll stick with the low-carb lifestyle forever. This is quite possibly the best thing you'll ever do for your health. But the benefits won't last if you revert back to your old ways. If you stray, so will your body—it will go back to its prior state after about two months (about the same amount of time it takes a fully fit body to go from physical greatness to being totally out of shape). So before you execute the steps, ask yourself why you want to change. Be honest and write those reasons down. Then go ahead and take a selfie—the "before" picture of yourself. Mark on the calendar the day you will begin. That is the day you accept the challenge, knowing that you are making a real commitment to your health. This is not a short-term diet. This is a *whole life* change. And your body—and future—will love it.

Believe in yourself, even when others don't: You will encounter people who support your new lifestyle and others who will try to sabotage it. Some individuals will express curiosity at your new dietary choices and some will mock you or tell you that you're ill-informed, stupid, or downright crazy. They may even include your best friends and family members. Be prepared to face these challenging, often embarrassing, encounters. When you turn down your cousin's homemade pie at Christmas, let alone all the other things on the dinner table, know what you're going to say: "I've been on this new diet and feel too fantastic to divert; unfortunately, certain ingredients are not allowed. Do you want to hear more about it?" With discourse and information comes understanding. Though you will face some people who will remain skeptics, don't let them bring you down. They may act appalled that you choose not to have a slice of pizza or a sandwich with them for lunch, but stay strong and comfortable in your decision. Your goal is to be the healthiest you can be. People will always judge. I bet that once you get used to fending off the naysayers and explaining yourself, soon they will follow in your footsteps.

Remember, it's a good idea to check with your doctor about beginning this new protocol, especially if you have any health issues for which you take prescribed medication. This is important if you're going to opt for the twenty-four-hour fast at the start. As you commence this new lifestyle, you will achieve the following important goals:

- Introduce a new way of nourishing your body, including your microbiome and brain, through the foods you eat
- Support the structure and function of your whole body through the right blend of supplements, probiotics included
- Add complementary strategies to the plan by focusing on more physical movement, restful sleep, attention to your emotional self and self-care, and cleaning up your physical environment

You know the rules. You know your goals. And you know the data that supports both. Now you are ready. Let's get to Step 1.

CHAPTER 5

Step 1—Edit Your Diet and Pill-Popping

WHAT EXACTLY IS ALLOWED ON this diet? The menu plan and recipes in Part III will help you follow the protocol, but let me give you a cheat sheet to guide you in shopping for and planning your meals. I'll also give you guidance on how to choose the right supplements to complement the diet and how to avoid certain medications if possible.

THE YES/NO KITCHEN CLEAN-OUT

The Grain Brain Whole Life Plan calls for the main dish to be mostly fibrous, colorful, nutrient-dense whole fruits and vegetables that grow above ground, with protein as a side dish. I cannot stress this enough: A low-carb diet is not all about eating copious amounts of meat and other sources of protein. To the contrary, a low-carb plate features a sizeable portion of vegetables (three-quarters of your plate) and just 3 to 4 ounces (about 100g) of protein (and no more than 8 ounces – 225g – of protein total in a day). You'll get your fats from those found naturally in protein, from ingredients used to prepare your meals such as butter and olive oil, and from nuts and seeds. The beauty of this diet is that you don't have to worry about portion control. If you follow these guidelines, your natural appetite-control systems kick into gear and you eat the right amount for your body and energy needs.

The Villains ("NO")

As you prepare for this new way of eating, one of the first things to do is eliminate items that you'll no longer be consuming. Start by removing the following:

All sources of gluten, including whole-grain and whole-wheat forms of bread, noodles, pastas, pastries, baked goods, and cereals. The following ingredients can also hide gluten and should be banished from your kitchen (and check labels to make sure other products don't contain these):

Avena sativa (a form of oats)

baked beans (canned)

barley

beer

blue cheeses

breaded foods

brown rice syrup

broths/stocks (commercially prepared)

bulgur

caramel color (frequently made from barley)

cereals

chocolate milk (commercially prepared)

cold cuts

couscous

cyclodextrin

dextrin

egg substitute

energy bars (unless certified gluten-free)

farina

fermented grain extract

flavored coffees and teas

french fries (often dusted with flour before freezing)

fried vegetables/tempura

fruit fillings and puddings

gravy

hot dogs

hydrolysate

hydrolyzed malt extract

ice cream

imitation crabmeat, bacon, etc.

instant hot drinks

kamut

ketchup

malt/malt flavoring

malt vinegar

maltodextrin

marinades

matzo

mayonnaise (unless certified gluten-free)

meatballs, meat loaf

modified food starch

natural flavoring

nondairy creamer

oat bran (unless certified gluten-free)

oats (unless certified gluten-free)

phytosphingosine extract

processed cheese (e.g., Kraft cheese slices)

roasted nuts

root beer

rye

salad dressings

sausage

seitan

semolina

soups

soy protein

soy sauces and teriyaki sauces

spelt

tabbouleh

trail mix

triticale

Triticum aestivum (a form of wheat)

Triticum vulgare (a form of wheat)

vegetable protein (hydrolyzed vegetable protein and textured vegetable protein)

veggie burgers

vodka

wheat

wheat germ

wine coolers

yeast extract

Be extra cautious of foods labeled “gluten-free,” “free of gluten,” “no gluten,” or “GF certified.” Although the FDA issued a regulation in August 2013 to define the term “gluten” for food labeling (the gluten limit for foods that carry any gluten-free label has to be less than 20 parts per million in America), it’s the responsibility of the manufacturers to comply and be accountable for using the claim truthfully. Some of the foods listed above, such as energy bars and mayonnaise, do come certified gluten-free and quality brands exist today. Do your homework, though. A gluten-free energy bar, for example, may contain lots of sugars and artificial ingredients that you want to avoid. Just because a food is labeled "gluten-free" and "organic" doesn't mean it lives up to my guidelines. And such products can derail your best efforts to put my protocol into practice and reap the health benefits.

Many foods marketed as being gluten-free never contained gluten to begin with (such as water, fruits, vegetables, eggs). But the term "gluten-free" does not indicate that a food is organic, low-carb, or healthy. In fact, food manufacturers use this term on products that have been processed so that their gluten has been replaced by another ingredient such as cornstarch, cornmeal, rice starch, potato starch, or tapioca starch, any of which can be equally as offensive. These processed starches can be allergenic and pro-inflammatory.

All forms of processed carbs, sugar, and starch:

agave
cakes
candy
chips
cookies
corn syrup
crackers
doughnuts
dried fruit
energy bars
fried foods
frozen yogurt/sherbet
honey
jams/jellies/preserves
juices
maple syrup
muffins
pastries
pizza dough
soft drinks
sports drinks
sugar (white and brown)
sugary snacks

Most starchy vegetables and those that grow below the ground:

beetroot	sweet potatoes
peas	sweetcorn
potatoes	yams

Packaged foods labeled "fat-free" or "low-fat": unless they are authentically fat-free or low-fat and within the protocol, such as water, mustard, and balsamic vinegar.

Margarine, vegetable shortening, trans fats (hydrogenated and partially hydrogenated oils), any commercial brand of cooking oil (soybean, corn, cottonseed, rapeseed, groundnut, safflower, grape seed, sunflower, rice bran, and wheat germ oils): even if they are organic. People often mistake vegetable oils as being derived from vegetables. They are not. The term is incredibly misleading, a relic from the days when food manufacturers needed to distinguish these fats from animal fats. These oils typically come from grains such as corn, seeds, or other plants such as soybeans. And they have been highly refined and chemically altered. The majority of Americans today get their fat from these oils, which are high in pro-inflammatory omega-6 fats as opposed to anti-inflammatory omega-3 fats. Do not consume them.

Non-fermented soy (e.g., tofu and soy milk) and processed foods made with soy: Look for "soy protein isolate" in the list of ingredients; avoid soy cheese, soy burgers, soy hot dogs, soy nuggets, soy ice cream, soy yogurt. Note: Although some naturally brewed soy sauces are technically gluten-free, many commercial brands have trace amounts of gluten. If you need to use soy sauce in your cooking, use tamari soy sauce made with 100 percent soybeans and no wheat.

The Heroes ("YES")

First things first: Remember to choose organic wherever possible and non-GMO foods, which will help you steer clear of gut-busting, fattening glyphosate. Choose antibiotic-free, grass-fed, 100 percent organic beef and poultry. This is key, because "grass-fed" doesn't necessarily mean "organic." When buying poultry, seek pastured meats that are also certified organic. This means the poultry is raised right on top of living grasses where they can eat all the various grasses, plants, insects, and so on that they can find in addition to their feed. When buying fish, choose wild, which often have lower levels of toxins than farmed.

Beware of the term "natural." The FDA has not fully defined the word, other than to say that it can be used on foods that do not contain added color, artificial flavors, or synthetic substances. But note that "natural" does not mean "organic," and it doesn't necessarily mean a food is healthy. It could still be loaded with sugar, for example. When you see this term, make sure to read the ingredient list.

Vegetables:

alfalfa sprouts

artichoke

asparagus

bok choy

broccoli

brussels sprouts

cabbage

cauliflower

celery
chard
fennel
garlic
ginger
green beans
jicama
kale
leafy greens and lettuces
leeks
mushrooms
onions
parsley
radishes
shallots
spinach
spring onions
turnips
water chestnuts
watercress

Low-sugar fruits:

aubergines
avocados
bell peppers
courgettes
cucumbers
lemons
limes
pumpkin
squash
tomatoes

Fermented foods:

fermented meat, fish, and eggs
kefir
kimchi
live-cultured yogurt
pickled fruits and vegetables
sauerkraut

Healthy fats:

almond milk

avocado oil

cheese (except for blue cheeses)

coconut oil (see note below)

coconuts

extra-virgin olive oil

ghee

grass-fed tallow and organic or pasture-fed butter

medium-chain triglyceride (MCT) oil (usually derived from coconut and palm kernel oils)

nuts and nut butters

olives

seeds (flaxseed, sunflower seeds, pumpkin seeds, sesame seeds, chia seeds)

sesame oil

A note about coconut oil: This superfuel for the brain also reduces inflammation. It's known in the scientific literature as helping to prevent and treat neurodegenerative disease. Use more of it when preparing meals. Coconut oil is heat-stable, so if you are cooking at high temperatures, use this instead of olive oil. (And if you don't like cooking with it, then you can take a teaspoon or two straight, as if it were a supplement—see page 105.) Coconut oil is also a great source of medium-chain triglycerides (MCT), an excellent form of saturated fatty acid. You can also add it to coffee and tea.

Proteins:

grass-fed meat, fowl, poultry, and pork (beef, lamb, liver, bison, chicken, turkey, duck, ostrich, veal)

shellfish and mollusks (shrimp, crab, lobster, mussels, clams, oysters)

whole eggs
wild fish (salmon, black cod, mahimahi, grouper, herring, trout, sardines)
wild game

Herbs, seasonings, and condiments:
cultured condiments (lacto-fermented mayonnaise, mustard, horseradish, hot sauce, relish, salsa, guacamole, salad dressing, and fruit chutney)
horseradish
mustard
salsas, if they are free of gluten, wheat, soy, and sugar
tapenade

Note: Sour cream, while technically a fermented dairy product, tends to lose its probiotic power during processing. Some manufacturers, however, add live cultures at the end of the process; look for brands that indicate this on the label ("with added live cultures").

Other foods that can be consumed occasionally (small amounts once a day or, ideally, just a couple times a week):
carrots
cow's milk and cream: Use sparingly in recipes, coffee, and tea
legumes (beans, lentils, peas): With the exception that chickpeas and hummus are fine, as long as they are organic. Watch out for commercially made hummus that's loaded with additives and inorganic ingredients. Classic hummus is simply chickpeas, tahini, olive oil, lemon juice, garlic, salt, and pepper

non-gluten grains:
- amaranth
- buckwheat
- millet
- quinoa
- rice (brown, white, wild)
- sorghum
- teff

parsnips

A note about oats: Make sure any oats you buy are truly gluten-free; some come from plants that process wheat products, causing contamination. I generally recommend limiting non-gluten grains because when processed for human consumption (such as milling whole oats and preparing rice for packaging), their physical structure can change, and this may increase the risk of an inflammatory reaction.

Sweeteners: natural stevia and chocolate (at least 75 percent cacao)

Whole sweet fruit: Berries are best; be extra cautious of sugary fruits such as apricots, mangos, melons, papayas, plums (or prunes), and pineapples

Label Lookout

Organic certification means that a food was produced without synthetic pesticides, genetically modified organisms (GMOs), or fertilizers made from petroleum. When it comes to organic meats and dairy products, it also means that they are from animals fed organic, vegetarian feed, are not treated with antibiotics or hormones, and are provided access to the outdoors. If the product was made with 100 percent organic ingredients, then it can say "100% organic." Just the word "organic" means the food was made with at least 95 percent organic ingredients.

"Made with organic ingredients" signifies that the product was made with a minimum of 70 percent organic ingredients, with restrictions on the remaining 30 percent, including no GMOs. As with the term "natural," "organic" doesn't imply healthy. Many organic junk foods line the shelves in supermarkets today, including candy and baked goods that are anything but healthy. When in doubt, scrutinize the ingredient list. It's the best way to know.

A note about buying produce: There's nothing more frustrating than splurging on fresh produce that wilts and begins to mold the moment you get home. Plan which fruits and vegetables you intend to use in the coming days based on the meal plan and purchase them on an as-needed basis, unless you're going to stock up on flash-frozen varieties. Flash-frozen fruits and vegetables are fine to buy as long as you choose organic or non-GMO verified. Three more tips:

- Avoid damaged, discolored, dull-looking fruits and vegetables. Ask your grocer what just came in and what's local. Stick with what's in season if you're buying fresh produce. If you crave berries, but the "fresh" ones have been shipped from thousands of miles away, opt for organic flash-frozen ones instead. These will have been picked during the peak of their ripeness, thus retaining their nutrients.
- Brighter means better: The brighter the colors you see, the more nutrients the fruit or vegetable contains. When you have choices in color, such as in bell peppers or onions, choose an array. Different colors impart different nutrients.

- High-risk crops for GMO: Papaya, courgettes, and yellow summer squash are often genetically modified foods, so look for non-GMO verified when buying these.

What to Drink?

The number one drink to have on hand at all times is filtered water, which is free of chemicals that can assault your microbiome. I urge you to buy a household water filter for all of your drinking and cooking purposes. There are a variety of water treatment technologies available today, from simple filtration pitchers you fill manually to under-the-sink contraptions or units designed to filter the water coming into your home from its source. I'm a big fan of the systems that employ reverse osmosis and carbon filters, so check those out if possible. It's up to you to decide which system best suits your circumstances and budget. Make sure the filter you buy removes fluoride, chlorine, and other potential contaminants. It's important that, with whichever filter you choose, you follow the manufacturer's directions to maintain it so that it continues to perform. As contaminants build up, a filter will become less effective, and it can then start to release chemicals back into your filtered water.

Other beverages that are allowable include coffee, tea, and wine (preferably red) in moderation. These drinks contain compounds that support gut and brain health. Just be sure not to overdo it. Coffee and tea contain caffeine, which can interfere with your sleep (unless you choose decaffeinated). In addition to green tea, which contains compounds known to fire up that Nrf2 pathway we discussed, I highly recommend trying kombucha tea. This is a form of fermented black or green tea that has been used for centuries. Fizzy and often served chilled, it's believed to help increase energy, and it may even help you lose weight. Note that wine should be limited to one glass for women and two for men per day.

Build and Maintain a Herb and Spice Collection

There's no better way to liven up meals than to add a dash of spice or a pinch of fresh herbs. Culinary herbs and spices can transform a dish from drab to fab. Although some do get expensive, you don't have to run out and spend a fortune in one fell swoop to create a spice rack worthy of a cooking magazine. Build it up over time. Here's a list of items you'll want to start collecting and experimenting with in your dishes. Choose garden-fresh organic herbs and non-irradiated herbs and spices wherever possible. You can start by purchasing 25g of each of the herbs and spices you want to try; for items you buy dried, store them in their original containers or transfer them to glass bottles that you can label. For fresh varieties, store them in the refrigerator and use them quickly.

- allspice
- basil
- bay leaves
- black pepper
- cayenne pepper
- chilli powder
- chives
- cilantro (coriander)
- cinnamon

- cloves
- cumin
- curry powder (red and yellow)
- dill
- garlic (powder and fresh cloves)
- ground ginger (and gingerroot)
- mint
- mustard seeds (black and yellow)
- nutmeg
- oregano
- paprika
- parsley
- red chilli flakes
- rosemary
- saffron
- sage
- savory
- sea salt
- tarragon
- thyme
- turmeric
- vanilla pods

Restock Your Pantry

If you followed the kitchen clean-out lists, chances are your pantry might be feeling lonely. You likely had to dump a lot of villains. So what goes in there now, besides your oils and vinegars?

- almond flour
- broth (beef, chicken, and vegetable)
- canned fish (salmon, tuna, anchovies)
- canned tomatoes (including paste)
- canned vegetables
- cocoa powder (at least 75 percent cacao)
- dill pickles
- hot sauces
- nuts and seeds

Now that you've edited your kitchen, it's time to edit your medicine cabinet.

PILLS TO POP — OR NOT

Every week it seems that we hear something in the media regarding the use of supplements. One day it's reported that certain vitamins are good for us and will extend our life; the next, we read that some can increase our risk for certain diseases, including dementia. While it's true that vitamins and supplements should never be used as an insurance policy against lapses in our diet, there's a time and place for certain products. And there's a difference between megadosing on multivitamins and adding natural supplements of nutrients that the body won't otherwise obtain easily from the diet.

There are many manufacturers of supplements today, and two formulas created for the same effect can have a different combination of ingredients as well as different dosages. Do your homework to find the best, highest-quality supplements that don't contain fillers, whether you're buying general supplements or probiotics. A good way to know which brands are best is to speak with the supervisor of the supplement and probiotic section at your health food store or retailer. These people tend to be knowledgeable about the top products and are not representatives of any particular company, so their advice is not biased. Supplements, probiotics included, are not regulated by the FDA like pharmaceuticals are, so you don't want to end up with a brand whose claims don't match the actual ingredients.

If a package of "high-potency" probiotic supplements with ten probiotic strains, for example, is marketed as delivering "50 billion live cultures per capsule," by the time you purchase the product, you may not be consuming that amount. The freshness and viability of packaged probiotic strains decline over time, even under ideal storage conditions. This is why it's important to know what you're getting and to ask for the superior brand with a reputable track record. Purchase your probiotics in smaller quantities and buy more frequently rather than buying the supersize value packs.

Note: If you currently take any prescription medication, it's important that you talk with your physician before starting any supplement program. You can take most of these supplements whenever it's convenient for you. Most do not have to be taken with food. The two exceptions: Probiotics should be taken on an empty stomach, and acacia gum, a prebiotic fiber I highly recommend that's now widely available in health food stores around the country, should be taken before an evening meal (more details follow).

It's usually best to take the supplements at the same time every day so you don't forget, which for many people is in the morning, before leaving home. Only one of my suggested supplements, turmeric, should be taken twice daily; have one dose in the morning and another in the evening. On page 112, I've created a cheat sheet for you to use that lists all the supplements and probiotics and their recommended dosages.

Following are my recommendations.

General Supplements to Consider

DHA: Docosahexaenoic acid (DHA) is a hero in the supplement kingdom and one of the most well-documented darlings in protecting the brain. DHA is an omega-3 fatty acid that makes up more than 90 percent of the omega-3 fats in the brain. Fifty percent of the weight of a neuron's membrane is composed of DHA, and it's a key component in heart tissue. DHA deficiency is seen in several disorders, including dementia and anxiety. The richest source of DHA in nature is human breast milk, which explains why breast-feeding is continually touted as important for neurologic health. DHA is also now added to formula as well as to hundreds of food products. Take 1,000 mg daily. It's fine to buy DHA that comes in combination with EPA (eicosapentaenoic acid), and it doesn't matter whether it's derived from fish oil or algae.

Coconut oil: As mentioned on page 95, if you don't cook often with this oil or use it in coffee and tea, enjoy its benefits by consuming a teaspoon or two once a day.

Turmeric: A member of the ginger family, turmeric is the seasoning that gives curry powder its yellow color. It's long been known for its anti-inflammatory, antioxidant, and anti-apoptotic properties—it reduces cell suicide or apoptosis. Turmeric is being studied today for its potential applications in neurology. Research shows that it can enhance the growth of new brain cells, as well as increase DHA levels in the brain. In some people, turmeric can even rival the antidepressant effects of Prozac. It's been used for thousands of years in Chinese and Indian medicine as a natural remedy for a variety of ailments. Curcumin, the most active constituent of turmeric, activates genes to produce a vast array of antioxidants that serve to protect our precious mitochondria. It also improves glucose metabolism, which helps maintain a healthy balance of gut bacteria. If you're not eating a lot of curry dishes, I recommend a supplement of 500 mg twice daily.

Alpha-lipoic acid: This fatty acid is found inside every cell in the body, where it's needed to produce energy for the body's functions. It crosses the blood-brain barrier and acts as a powerful antioxidant in the brain. Scientists are now studying it as a potential treatment for strokes and other brain conditions, such as dementia, that involve free-radical damage. Although the body can produce adequate supplies of this fatty acid, it's best to supplement to ensure you're getting enough. Aim for 300 to 500 mg daily.

Coffee fruit extract: This is one of the most exciting additions to my supplement regimen. This extract, which contains very little caffeine, has recently been shown to increase blood levels of a protein called "brain-derived neurotrophic factor," or BDNF. I can't emphasize enough how important BDNF is, not only to keeping the brain healthy and maintaining its resistance to damage, but also to triggering the growth of new brain cells and increasing the connections between them. Study after study shows a relationship between levels of BDNF and risk for developing Alzheimer's disease. In a seminal 2014 study published in the prestigious *Journal of the American Medical Association*, researchers at Boston University found that in a group of more than 2,100 elderly people followed for ten years, 140 of them developed dementia. Those with the highest levels of BDNF in their blood had less than half the risk for dementia as compared to those who had the lowest levels of BDNF. Low levels of BDNF are documented in people with Alzheimer's as well as in people with obesity and depression. Look for whole coffee fruit concentrate, and take 100 mg daily. A single dosage of whole coffee fruit extract has been shown to *double* blood levels of BDNF during the first hour after consumption.

Vitamin D: This is technically a hormone, not a vitamin. By definition, vitamins cannot be produced by the body. But vitamin D is produced in the skin upon exposure to ultraviolet (UV) radiation from the sun. Although most people associate it with bone health and calcium levels, vitamin D has far-reaching effects on the body and especially on the brain. We know there are receptors for vitamin D throughout the entire central nervous system; in fact, researchers have identified approximately 3,000 binding sites on the human genome for vitamin D. We also know that vitamin D helps regulate the enzymes in the brain and cerebrospinal fluid that are involved in manufacturing neurotransmitters and stimulating nerve growth. Both animal and human studies have indicated that vitamin D protects neurons from the damaging effects of free radicals and reduces inflammation. As previously noted, vitamin D is also associated with longer telomeres. And here's a most important fact: Vitamin D performs all of these tasks through its relationship with the gut bacteria. In 2010 we found out that gut bacteria interact with our vitamin D receptors, controlling them to either increase their activity or turn it down.

As mentioned on page 75, I encourage you to get your vitamin D levels tested and have your doctor help you find your optimal dose. For adults, I generally recommend starting with 5,000 IU of vitamin D daily. Some people need more and others less. It's important to have your doctor track your vitamin D levels until you land on a dosage that will keep you in the upper range of normal on the blood test.

Supplements to Support Gut Health

The two secrets to improving the composition and function of your gut bacteria are prebiotics and probiotics.

Prebiotics

Prebiotics, the ingredients that gut bacteria love to eat to fuel their growth and activity, can easily be ingested through certain foods. To qualify as prebiotics, they must have three characteristics. First and foremost, they must be nondigestible, meaning they pass through the stomach without being broken down by either gastric acids or enzymes. Second, they have to be able to be fermented or metabolized by the intestinal bacteria. And third, this activity has to confer health benefits. Prebiotic dietary fiber, for example, meets all these requirements, and its effects on the growth of healthy bacteria in the gut may well be the reason it's anticancer, anti-diabetes, anti-dementia, and pro-weight loss.

By and large, we don't get anywhere near enough prebiotics. I recommend aiming for at least 12 grams daily, from either real foods, a supplement, or a combination thereof. Again, this is one of the most important steps you can take to nurture the health and function of your good gut bacteria and open the door for a healthy future for yourself. Below is the list of top food sources of natural prebiotics.

- acacia gum (or gum arabic)
- asparagus
- chicory root
- dandelion greens
- garlic
- leeks
- onions
- sunchokes (Jerusalem artichokes)

While some of these may be a bit unfamiliar, my menu plan will show you how to make use of them and get plenty of prebiotic fiber into your diet every day. Health food stores now also carry powdered prebiotic fiber products that you can simply mix with water. These products, which are often derived from acacia gum, provide a convenient source of concentrated prebiotic fiber that will nurture your gut bacteria. Acacia gum has been extensively studied. It has been found to have a significant impact on weight loss. One recent study showed a dramatic reduction in both body mass index and body fat percentage among healthy adult women taking acacia gum as a nutritional supplement. The FDA considers acacia gum one of the safest, best-tolerated dietary fibers; it doesn't increase the risk of bloating, abdominal cramps, or diarrhea.

So if you are looking for a prebiotic fiber supplement, look for acacia gum. All you need is a level tablespoon or two a day in any beverage—fifteen to thirty minutes before the evening meal is ideal. While 12 grams of prebiotic fiber a day is a great target, it may take a week or two to be able to tolerate that much—you may experience some gas. You can start with just 1 tablespoon of acacia fiber daily and build up to 2 tablespoons daily.

When choosing a prebiotic supplement, look for:

- certified organic, Non-GMO Project verified, vegan, gluten-free labeling
- products that are free of psyllium, soy, and sugar
- no artificial colors, sweeteners, or flavors

Probiotics

As with prebiotics, you can get your probiotics through food and supplements. In terms of food, I recommend keeping the following in your kitchen:

- Live-cultured yogurt. The dairy section has gotten crowded. There are lots of options today when it comes to yogurt, but you have to be careful about what you are buying. Many yogurts—both Greek-style and regular—are loaded with added sugar, artificial sweeteners, and artificial flavors. Read the labels. If you are sensitive to dairy, try coconut milk yogurt. It is an excellent way to get plenty of gut-promoting enzymes and probiotics into your diet.

- Kefir. This fermented milk product is similar to yogurt. It's a unique combination of kefir grains (a symbiotic culture of yeast and bacteria) and goat's milk that's high in lactobacilli and bifidobacteria, two of the most studied probiotics in the gut. Kefir is also rich in antioxidants. If you are sensitive to dairy or are lactose-intolerant, coconut milk kefir is equally delicious and beneficial.
- Sauerkraut. This fermented cabbage fuels healthy gut bacteria and provides choline, a chemical needed for the proper transmission of nerve impulses from the brain through the central nervous system.
- Pickles. I believe pregnant women crave pickles for a reason. Pickles are one of the most basic and beloved natural probiotics. For many, pickles can be your gateway food to other, more exotic fermented foods.
- Pickled fruits and vegetables. Pickling fruits and veggies, such as carrot sticks and green beans, transforms the ordinary into the extraordinary. Whether you do this yourself or buy pickled produce, keep in mind that only unpasteurized foods pickled in brine, not vinegar, have probiotic benefits.
- Cultured condiments. You can buy or make your own lacto-fermented mayonnaise, mustard, horseradish, hot sauce, relish,

salsa, guacamole, salad dressing, and fruit chutney. Remember to look for sour cream with added live cultures.

- Fermented meat, fish, and eggs. See my website, www.DrPerlmutter.com, for brand ideas and recipes for these. It's best to make these on your own rather than buying commercially made products, which are often processed with other ingredients you don't want.

The number of probiotic supplements available today can be overwhelming. Thousands of different species of bacteria make up the human microbiome, but I have a few gems to recommend:

Lactobacillus plantarum
Lactobacillus acidophilus
Lactobacillus brevis
Bifidobacterium lactis
Bifidobacterium longum

Most probiotic products contain several strains, and I encourage you to seek a probiotic supplement that contains at least ten different strains, with as many of the above-mentioned species as possible. Different strains provide different benefits, but these are the ones that will best support brain health by:

- fortifying the intestinal lining and reducing gut permeability
- reducing LPS, the inflammatory molecule that can be dangerous if it reaches the bloodstream
- increasing BDNF (brain-derived neurotrophic factor), which is fondly known as the brain's "growth hormone"

- sustaining an overall balance of bacteria to crowd out any potentially rogue colonies

If you're wanting to lose weight, I suggest looking for the following species in addition to those above:

Lactobacillus gasseri
Lactobacillus rhamnosus

For those with mood issues, including depression, look for:

Lactobacillus helveticus
Bifidobacterium longum

Remember, plan to take your probiotics on an empty stomach, and aim to take them at least thirty minutes before a meal.

THE SUPPLEMENT CHEAT SHEET

Name	Amount	Frequency
DHA	1,000 mg	daily
Coconut oil	1–2 teaspoons	daily (if not using in cooking/coffee/tea)
Turmeric	500 mg	twice daily
ALA	300–500 mg	daily
Coffee fruit extract	100 mg	daily
Vitamin D	5,000 IU	daily
Prebiotic fiber	12 g	daily (15–30 minutes before dinner)
Probiotics	1 multi-strain capsule	daily (at least 30 minutes before a meal)

Pills that Should Give You Pause Before Popping

The vast majority of Americans take a medication of some sort daily, whether it's prescribed or over the counter. Nearly three in five American adults take a prescription drug; in 2015, the *Journal of the American Medical Association* published findings that the prevalence of prescription drug use among people twenty and older had risen to 59 percent in 2012 from 51 percent just a dozen years earlier. And the percentage of people taking five or more prescription drugs nearly doubled during the same time period. It went from 8 percent to 15 percent.

Among the most commonly used drugs are the ones that increase the risk of brain ailments: statins. I've made my case about statins in the past and I briefly outline my main lesson below. But statins aren't the only problem. I highly recommend that you take inventory of your medicine cabinet and aim to reduce the number of drugs you take unless they are absolutely necessary to treat a condition. (Obviously, speak with your doctor should you consider stopping any medication specifically prescribed for you.) Following are the worst offenders:

Statins: Cholesterol-lowering statins are now being sold as a way to reduce overall levels of inflammation. But new research reveals that these powerful chemicals may decrease brain function and increase risk for diabetes, heart disease, impaired cognitive function, and depression. The reason is simple: The body, and especially the brain, needs cholesterol to thrive. What's more, cholesterol is involved in cell membrane structure and support, hormone synthesis, and vitamin D production. Reams of scientific data show time and time again that extremely low cholesterol levels are linked to depression, memory loss, and even violence to oneself and others.

Acid-reflux drugs (proton pump inhibitors): An estimated 15 million Americans use proton pump inhibitors (PPIs) for gastroesophageal reflux disease, or GERD. These drugs are sold by prescription and over the counter under a variety of brand names, including Nexium, Prilosec, Protonix, and Prevacid. They block the production of stomach acid, something your body needs for normal digestion. In the past two years, the negative effects of these drugs have been shown in widely reported studies. Not only do they leave people vulnerable to nutritional deficiencies and infections, some of which can be life-threatening, but they also put people at greater risk of heart disease and chronic kidney failure. And they do a number on your gut bacteria. When researchers examined the diversity of microbes in stool samples of those taking two daily doses of proton pump inhibitors, they documented dramatic changes after just one week of treatment. These drugs can effectively ruin the integrity of your digestive system by dramatically changing your gut bacteria.

Acetaminophen: Nearly a quarter of American adults (about 52 million people) use a medicine containing acetaminophen (brand name Tylenol) each week for aches, pains, and fever. It's also the most common drug ingredient in the United States, found in more than 600 medicines. But it's not as benign as we've been led to believe. It's now been shown to be ineffective for osteoarthritis pain, the very condition for which it's heavily marketed. What's more, new research shows that it compromises brain function, increasing the risk of making cognitive mistakes. Although early research had shown that acetaminophen not only affects physical pain, but also psychological pain, we now know the true nature of its effects thanks to a 2015 Ohio State University study revealing that acetaminophen blunts emotions, positive and negative. Participants who took acetaminophen felt less strong emotions when they were shown both pleasant and disturbing photos as compared to the controls who were given placebos.

Acetaminophen is also known to deplete one of the body's most vital antioxidants, glutathione, which helps to control oxidative damage and inflammation in the body and especially in the brain. And in another 2015 study, Danish scientists found that women who took acetaminophen during pregnancy were more likely to have children medicated for ADHD by age seven. Tylenol is often "prescribed" to pregnant women and regarded as "safe." I hope that thinking changes soon.

Nonsteroidal anti-inflammatories: Think ibuprofen (Advil) and naproxen (Aleve). Like Tylenol, these are hugely popular pain relievers and fever reducers—on any given day about 17 million people take them. These drugs work by reducing the amount of prostaglandins in the body, a family of chemicals produced by the cells that have several important functions. Prostaglandins promote the kind of short-term inflammation necessary for healing; they support the blood clotting function of platelets; and they protect the lining of the stomach from the damaging effects of acid. Because of these last two functions, NSAIDs can compromise that intestinal lining; their number one side effect is stomach bleeding, ulcers, and stomach upset. Research shows that they damage the small intestine and indeed harm the gut lining, thereby setting the stage for the very problem they are intended to address: inflammation.

Antibiotics: This one should be obvious. Antibiotics are anti-life. They kill bacteria, both the good guys and the bad. Almost all of us need to take a round of antibiotics at some point in our lives. The effects of exposure to antibiotics on gut bacteria can persist for months after treatment, and new research has concluded that even one course of antibiotics can change the microbiome for the rest of a person's life. Such changes can have far-reaching effects on the body if the balance of healthy bacteria is not restored.

This new knowledge follows mounting evidence that they also drive adverse changes in insulin sensitivity, glucose tolerance, and fat accumulation due to how they alter the gut bacteria. The drugs also tinker with our own physiology, changing how we metabolize carbohydrates and how the liver metabolizes fat and cholesterol. Dr. Brian S. Schwartz of the Johns Hopkins Bloomberg School of Public Health, who has studied these connections, has gone so far as to say, "Your BMI may be forever altered by the antibiotics you take as a child."

Scientists have been tracking the strong correlation between exposure to antibiotics and risk for weight gain and type 2 diabetes. Take a look at the following two maps. On the left we see antibiotic prescriptions per 1,000 people, and on the right we see rates of obesity listed by state. These two maps look strikingly similar.

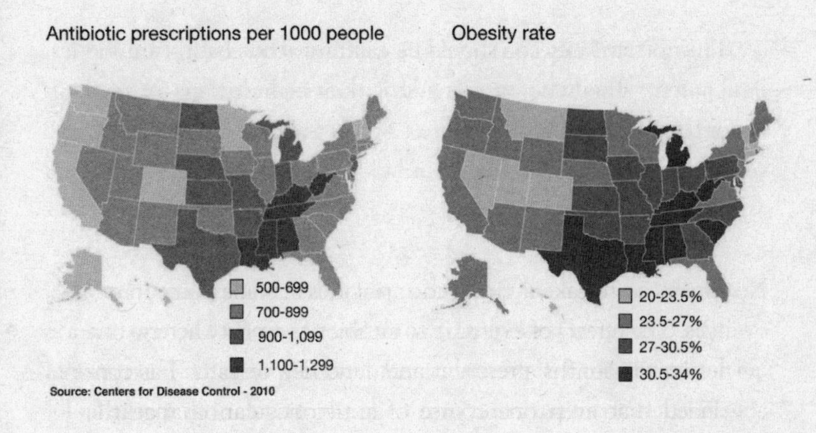

This next graphic shows antibiotics prescriptions per 1,000 people and prevalence of adult diabetes. Again we see a correlation. Also, remember that there is a notable relationship between obesity as well as adult diabetes and the risk for dementia. I think you can see the point I'm trying to reiterate: Our overuse of antibiotics is not only fueling our obesity and diabetes epidemics but also our increasing rates of dementia.

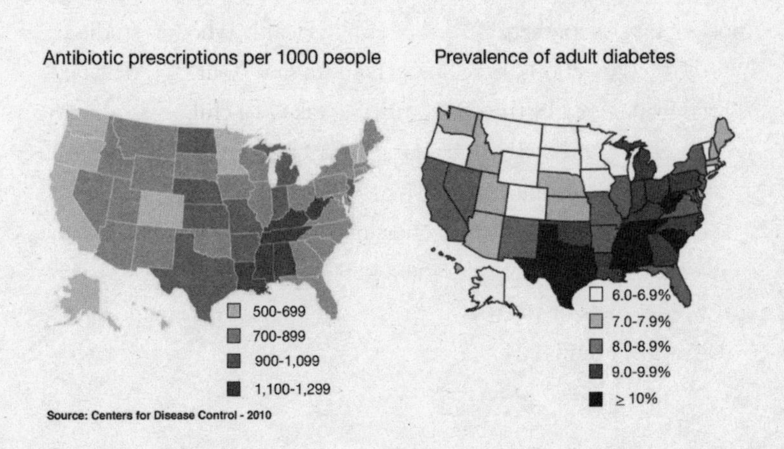

The bottom line: You should be cautious about taking antibiotics. And parents should be wary of asking their pediatricians for antibiotics for their children at the slightest sign of a sniffle. In Part III, I'll give you some guidelines for when antibiotics might be necessary for you and your family.

No doubt there is a time and place for medications, be they over-the-counter or prescribed. But we live in a world where we are too quick to medicate, self-prescribe, and depend upon pills. The national conversation about pain and pain relief in particular has sparked a lot of debate over painkiller abuse. In 2014 alone, more than 14,000 Americans died of overdoses involving prescription opioids, and each day more than 1,000 people enter emergency rooms for misusing those drugs. May we see the day when we can minimize medications and maximize our body's innate ability to heal itself. If you rely on medications, I encourage you to work with your health-care provider to find alternative methods for treating and managing your conditions. I do trust that if you follow this program you will experience a lessening of your symptoms, whether you need to continue treatment with drugs or not.

CHAPTER 6

Step 2—Add Your Support Strategies

In studies of populations around the world where the inhabitants live unusually long, healthy, productive lives well into their nineties and have less than half the rate of cancer as Americans, very little depression, and almost no dementia, there are a few common denominators. In these "Blue Zones," people maintain a positive outlook on life, have strong family and personal relationships, and feel part of a close-knit community. They are physically mobile in everyday life, consume fresh, local ingredients, and don't eat processed food. In the Blue Zone of Ikaria, Greece, fat accounts for more than 50 percent of Ikarians' daily calories, and more than half of their fat energy comes from olive oil. In Sardinia, Italy, where there's a community of mostly sheepherders, people spend their days walking, enjoying the company of others, and drinking local red wine with their meals. The famous Okinawan residents of Japan, many of whom live for about 110 years, base their diet on vegetables, drink mugwort sake, and remain physically active as they age. They also honor and celebrate their elders. The only Blue Zone in the United States is tucked just east of downtown Los Angeles, a city notoriously known for its smog and dense population. The community of Seventh-day Adventists in Loma Linda, California, near the big city, defies the belief that you have to live in a pristine, remote area to enjoy a long, disease-free life.

All of these people have mastered the tenets of healthy living without trying that hard or even knowing it. Self-care beyond the diet is sorely neglected in our society, yet factors like regular exercise, more sleep, and distraction-free time for self-reflection can make a huge difference in our health. The fast-paced world in which we live causes us to feel time-deprived and anxious, and during stress we turn to unhealthy habits that push us in the wrong direction and leave us ever more tired, uncreative, drug- and stimulant-dependent, and unsatisfied. Stress adds tremendous pressure on our biological systems, from the gut all the way up to the brain.

With that in mind, let me give you some non-diet-related strategies to boost overall health and prevent brain disease:

- Establish an exercise routine you can sustain
- Pay attention to pain, especially in your back and knees
- Make room for sleep
- Reduce stress and find calm in four simple ways
- Detoxify your physical environment

ESTABLISH AN EXERCISE ROUTINE YOU CAN SUSTAIN

Did you know that you can physically bulk up your brain with exercise and slash your risk for Alzheimer's disease in half?

The evidence is no longer anecdotal. Every week a new study emerges showing the neuroprotective benefits of exercise. Sedentarism appears to cause the brain to atrophy while simultaneously increasing the risk for Alzheimer's disease and other types of dementia. Being sedentary has even been shown to be twice as deadly as being obese.

In February 2016, a Finnish study found that being out of shape in middle age is linked to smaller brain volume down the road in later life. Gray matter is where all of your neurons are, so its volume can reflect brain health. Soon after, another study, conducted by researchers at four major research institutions in the United States, found that people with Alzheimer's disease or mild cognitive impairment (MCI, a precursor to Alzheimer's disease) experienced less gray matter shrinkage the more calories they burned through their exercise regimen. In other words, more physical activity meant more retained brain volume and less risk of disease. The researchers followed 876 adults over a thirty-year period and kept careful records of the amount and type of exercise that each participant reported. In addition, each of the individuals underwent a rigorous examination to determine brain function. Further, all of the participants had brain imaging done with a sophisticated MRI scanner. Those at the highest level of exercise activity experienced an incredible reduction of risk for Alzheimer's of 50 percent when compared to those who were more sedentary.

We live in a time when we have been fairly well trained to expect "magic bullets" to resolve our ailments. When it comes to Alzheimer's, none exists. Yet look at how you can safeguard your brain just by lacing up your sneakers and moving.

In addition to protecting the brain, exercise improves digestion, metabolism, elimination, immunity, complexion, body tone and strength, bone density, and circulation and heart health, and it helps us normalize weight. Just a twenty-five minute walk can cut your risk of dying prematurely by 30 percent; a brisk walk for the same time can add seven years to life. Physical movement is also an emotionally positive experience. It can increase self-worth and confidence and bring us greater energy. It can turn on our "smart genes," make us feel younger, stave off depression, and generally help us make healthier lifestyle choices, including what we're eating for dinner.

Regular aerobic exercise:

lowers the risk of death from all causes

lowers the risk of brain disorders

lowers the risk of depression (and can be used as a treatment for depression)

increases the quantity of BDNF, the brain's "growth hormone"

increases stamina, strength, flexibility, and coordination

increases blood circulation, oxygen supply to cells and tissues, and heart health

decreases food cravings, blood sugar levels, and risk for diabetes

decreases inflammation and risk for age-related disease, including cancer

increases sense of well-being and happiness

The last time you tried to get into shape, what happened? Did you succeed? Or did you keep up a routine for a few weeks, maybe right after New Year's, and then suddenly it was Memorial Day weekend and you didn't want to be seen in a bathing suit? Maybe you don't even remember when you fell off the wagon, but it happened.

The problem for anyone trying to get active is not so much the starting part as it is the *sustaining* part. The key is to figure out what it is that you love to do and that achieves the following: 1) stretches and strengthens your muscles, and 2) gets your blood flowing physically, increases your heart rate, and puts healthy demands on your cardiovascular system.

Check with your doctor before starting an exercise program if you have any personal conditions and/or use medication that might be a factor in this endeavor.

Stretching and strengthening is important because far too many people focus solely on cardio and skip the stretching and weight-bearing exercises. If you don't stretch and strengthen your muscles, you not only impair bone health and muscle mass, but you also run the risk of injury, which will prevent you from staying active. Starting on page 129, I've created a very basic strength-training routine you can do at home. I also invite you to go to www.DrPerlmutter.com, where I provide video clips of these exercises that I encourage you to do in addition to a cardio routine of your choice. These clips cover all the major muscle groups—arms and shoulders, chest, back, abdominals, and legs. Perform these exercises three or four times a week, and have a rest day in between each session.

Ideally, aim for a **minimum of twenty minutes of cardio work six days a week**. You'll want to get your heart rate up at least 50 percent above your resting baseline for a least fifteen of those twenty minutes. There are lots of different kinds of heart-rate monitors on the market; and various types of gym equipment, such as stationary bikes, elliptical machines, and treadmills, now include heart-rate monitors, too. Online calculators can also help you find your maximum and target heart rates so you know when you're in the zone, as well as when you're pushing the limits. At first you may not be able to hit any reasonable target heart rate or, if you can, you may not be able to sustain it for long. But you can build up to that target. I've often said to my patients, "If you can't run five miles, then at least walk to the mailbox." You have to start somewhere, and if walking to the mailbox represents your first steps toward improving your health, then so be it. Even people who are confined to a wheelchair can and absolutely should engage in aerobic exercise.

Golfing doesn't count: One frequent response I get when asking patients about their exercise program, likely because I live in South Florida, goes something like this: Well, Doctor, I play eighteen holes of golf three times a week." I have nothing against golf, but this doesn't constitute aerobic exercise—even if you're walking the course, which is painfully rare these days.

Create a realistic plan that you can maintain. For some, that might mean participating in group classes at a local gym; for others, spending more time gardening, taking up yoga and swimming, joining a competitive sports team in your community, power walking around the mall, or following a workout routine online or on TV. I have been a long-distance runner since high school, and recently I have been using the elliptical machine as well as a stationary bike and road bike for aerobics. I go hard on some days and easier on others. I suggest you do the same: Mix up the days you work out at a high intensity for shorter periods of time and then have those days when you go at a moderate pace for a longer time. The twenty-minute rule should be your minimum. You should be able to add more minutes as you gain strength and physical fitness. Build up to more intense workouts, too. Intensity can be increased through speed (e.g., running or pedaling faster), resistance (e.g., steeper climbs, heavier barbells), duration (e.g., longer intervals of giving it your all to the point you're out of breath), and range of motion (e.g., deeper lunges, lower bends).

The optimal amount of exercise time, in fact, to gain the most benefits is close to 450 minutes per week. That averages out to a little more than an hour a day, which may seem like a lot, but it's not when you consider that this amount reflects *cumulative* minutes of exercise. You don't have to max out your heart rate for an entire hour, either, but you do need to be moving your body physically for at least that amount of time in total on most days of the week. It's easier than you think. You might do a twenty-minute jog in the morning, run errands at lunch during which you walk briskly for another twenty minutes, and then spend twenty minutes before dinner tending to household chores that are physically demanding. If you skimp on exercise one day, that just means you go longer the next. It doesn't really matter how you allocate those 450 minutes over the course of the week. Try to be as consistent as possible with when you plan your formal exercise daily (e.g., twenty minutes of cardio every morning before taking a shower), but don't beat yourself up if you're not perfect every single day. There will be days when you'll need to move the timing of your usual routine and days when fitting in formal exercise isn't possible. Strive for progress, not perfection.

And you don't have to get too techy, despite the number of devices available to track physical parameters like heart rate. Just as pain can be your guide to let you know when you've overdone it (more on this coming up), how you feel in the moment can be as good as any high-tech gadgetry. Pay attention to your breathing and level of sweating. Does your breath get deep and rapid during your formal workouts? Do you break a sweat? Do your muscles start to burn a little during weight-bearing exercises and feel a tad sore the next day while resting? There is a difference between running and leisurely mowing the lawn, as well as between using 3-pound versus 8-pound weights.

We all think in pictures, and studies show that imagining yourself in the physical shape you want to be in can help you reach your fitness goals. Try to keep a vivid, realistic image in your mind. This will motivate you every step of the way as you work toward your personal picture of health. Think about what it means for you to have a fit, toned body. You'll be able to participate in life to its fullest and not be constrained by low energy and no strength. Visualize yourself engaged in various fun activities you want to try, including adventures and vacations that make physical demands on you. Consider activities done alone, with a group, or with your family.

Concentrate also on what you'll gain in terms of vigor, balance, coordination, flexibility, and mental sharpness (and mental toughness!). You're sleeping better, managing stress more easily, enjoying a faster metabolism, being more productive overall, and spending less time sick in bed with a cold or other illness. You know that you're doing the best you can to ward off disease. If you are dealing with any chronic conditions, you're managing them superbly and they are having less impact on you. You're feeling more accomplished both at work and at home—because you are! And you're experiencing stronger, more intimate relationships with your loved ones.

How to Create Your Own Gym for Strength Training (No Trainer Required!)

You don't need much in the way of equipment to get a great strength-training workout. In fact, you don't even have to belong to a traditional gym, hire a trainer, or spend money on machines and fancy gadgetry to generate resistance against your body. There's a lot you can do using your own body weight. Classic push-ups and sit-ups, for example, don't require anything but you and a floor. But to get a full-body strength-training workout, I suggest you obtain a couple of inexpensive free weights. You can buy them online or at a local sporting goods store. Choose weights that are comfortable to grip. Start on the light side (3- or 5-pound – 1 or 2kg – weights) and add more weight as you gain strength and want to challenge yourself further.

Although there are many muscles and muscle groups in the body to work on a regular basis, it helps to think in terms of upper body, lower body, and core:

- Upper body: shoulders, triceps, biceps, chest, and lats (latissimi dorsi muscles of the back)
- Lower body: thighs/quads and calves
- Core: midsection and abdominals

On the days you choose to do your strength training—again, which you should do three or four times a week—perform exercises from each of the three main areas. While it would be optimal to hit every muscle group during your sessions, if you're short on time, you can split up the areas you focus on. So, for instance, if you exercise your triceps, biceps, calves, and core on Monday, you can turn to your shoulders, chest, thighs, and more core on Wednesday. Some people like to include strength training in their daily exercise routine, which is also okay as long as you don't repeat the same muscle groups two days in a row. Give those muscles at least a day or two of rest in between.

I encourage you to include a little bit of core work in each strength-training session. Powering up your core muscles and maintaining their strength is fundamental to health—more so than having toned arms. Your core is largely responsible for keeping you active and able to perform everyday tasks—from getting out of bed to sitting on the toilet, dressing, standing, and walking—as well as for playing sports and engaging in activities like biking, tennis, and dancing. Having a strong core prevents back pain, provides stability and balance, propels your endurance, and supports good posture. You don't have to aim for ripped, washboard abs. Far from it. You just need to work your midsection and abdominals routinely to prevent weak and inflexible core muscles. In fact, a weak or inflexible core can hinder how well your legs and arms function, draining energy from your movement and impairing everyday activities.

There are dozens of different exercises that will work your upper body, your lower body, and your core. And many exercises will engage the core even though you're focusing on another body part. Following are some basic exercises for a strength-training workout. Most of these require free weights. Because many cardio routines also place heavy demands on various muscle groups, you'll find that some of your muscles will be quicker to tone and strengthen because they are getting more of a workout (e.g., taking a spin class will also work your quads and calves in ways that are equivalent to weight-bearing exercises; swimming will work your upper body and back).

Remember to go to my website, www.DrPerlmutter.com, where you can watch me demonstrate these very exercises. You may want to venture out and test other methods of resistance training, either by using formal gym equipment or by taking a group fitness class that is centered on strength training (e.g., Pilates, many forms of yoga, and gym classes geared specifically toward building muscle mass and strength). I usually go to the gym for strength training because I can access a wider array of tools.

Shoulders: Basic Lifts

Stand up straight with your feet hip distance apart, arms by your side. Hold one weight in each hand, keeping your shoulders down and chest open, and maintain good posture. Lift the weights out to the side to shoulder height (as if you're making a letter "T" with your body). As you lift, with palms facing down, squeeze your shoulder blades together, and then lower the weights. Complete three sets of twelve repetitions (lift and lower twelve times).

Try a variation: Instead of lifting your arms out to the side, lift them up in front of you with straight arms, palms facing down.

Triceps: The Triceps Extension

Hold a free weight with both hands overhead. Try to use a weight that's at least 5 pounds (2kg). Draw your shoulders down and back, and engage your core. While keeping your elbows pointed forward, bend the elbows and allow the weight to lower down behind your head. Then bring the weight back up and overhead by extending your arms. Keep your core and glutes engaged the entire time. Complete three sets of twenty repetitions.

Biceps: Basic Bicep Curl

Stand up straight with your feet hip distance apart, gripping a free weight in each hand. Your starting position is to have your hands down by your sides, palms facing forward. While keeping your elbows close to your torso and your upper arms stationary, lift your forearms up, curling the weights up while contracting your biceps. Complete three sets of twenty repetitions.

Chest: Classic Push-Ups

Lying face down on the floor, place your hands under your shoulders and tuck your toes underneath you. Push up into a plank. Hold for five seconds, then slowly lower down toward the floor, trying to achieve about a 90-degree bend in the elbows. Try not to collapse onto the floor, and repeat the push-up again into the plank position. Complete three sets of twelve push-ups.

Lats: The Wide Row

The best exercise to work these back muscles is to do pull-ups to a raised bar. But another way, using your free weights, is the following. Stand straight, chest up and back flat, while holding a pair of free weights in each hand in front of your thighs with a palms-down grip. Now, slightly bend your knees and lean forward, hinging at the waist. Continue to lean forward until your upper body is almost parallel to the floor. Let the weights hang straight down in front of your shins. With your head in a neutral position and your eyes focused on the floor in front of you, lift both weights straight up, bending at the elbows. This is a rowing-like motion but you're in a semi-squat. Don't alter the angles at your knees and hips and lower the weights back after a short pause. Complete three sets of twelve repetitions.

Thighs/Quads: Lunges

Stand up straight with your feet hip distance apart, and have a slight bend in your knees. Hold the free weights down at your sides. This is your starting position. Now, step forward with your right leg while maintaining your balance and squat down through your hips. Keep your torso straight and your head up. Don't let your knee extend out over your toes. Using your heel to drive you, push yourself back to the starting position. Repeat this motion with your left leg to complete the full rep. Do three sets of twelve repetitions.

Calves: Tippy-Toes

Stand up straight with your feet hip distance apart. Hold a free weight in each hand, with each weight hanging by your sides. Push up onto your tippy-toes, and hold there for five seconds. Return to the start. Complete three sets of twelve repetitions.

Core: The Classic Sit-Up

Sit on the floor with your knees bent and your heels touching the floor. Cross your arms on your chest, making an "X." Make sure to keep your shoulders dropped and relaxed to avoid tension in the neck. With your feet firmly on the ground, lay back as far as you're able before rising back up. Maybe you can get all the way to the floor, maybe not. Continue doing sit-ups for one minute, then take a thirty-second break. Repeat for five rounds.

Core: Bicycle Crunch

Begin in the same starting position as the sit-up (see above). Twisting gently, bring your left knee and right elbow toward one another. Return to the starting position. Complete the movement with the right knee and left elbow. Continue for two minutes, then take a thirty-second break. Repeat for five rounds.

* * *

Take the time to write down your reasons for making these important fitness changes in your life. Rather than stating, "I want a flatter stomach and toned arms," go for more meaningful, purposeful goals like "I want to spend more quality time with my family rather than constantly dealing with my chronic pain," "I want to feel stronger and live longer," or "I want to do everything I can to prevent the Alzheimer's disease that my mother has." Think big picture and be bold and brave with your goals.

PAY ATTENTION TO PAIN, ESPECIALLY IN YOUR BACK AND KNEES

I cannot emphasize enough the importance of not neglecting two parts of your body that are critical to staying mobile and therefore reducing your risk of disease: your lower back and knees. Let's cover the lower back first.

The numbers are staggering. After colds and influenza, back pain is the second most common reason Americans see their doctor, and it's the most common cause of job-related disability. Back pain is the second most common neurological ailment in the United States—only headaches are more common. And it's the third most common reason for emergency room visits. At some point, more than 90 percent of American adults will experience severe lower back pain that adversely affects their quality of life. It's estimated that lower back pain costs the American economy $50 to $100 billion annually.

In my thirty-odd years of practice I saw lower back pain routinely. Early on in my practice, many of these patients would be referred to neurosurgeons because in those days, it was believed that most lower back pain was caused by "ruptured disks." We now know that this was completely off base and that, by and large, lower back pain is only rarely caused by disk problems. It is almost always caused by damage in the soft tissues—meaning muscles, tendons, and ligaments.

Although many things can cause back pain, from strained muscles to cancer, I do want to highlight one condition in particular that is extremely common but under-recognized: piriformis syndrome. Your piriformis (from the Latin word meaning "pear-shaped") is a narrow muscle deep in each buttock. These muscles lie close to the sciatic nerve, so when the piriformis muscle spasms or is irritated, it can aggravate the sciatic nerve and trigger pain to shoot right down the leg from the buttock, as if it were coming from a ruptured disk. Patients are told that they have "disk disease" because of pain going down the leg, called sciatica. They may also experience numbness and tingling along the back of the leg and into the foot due to the irritated sciatic nerve.

It's hard to imagine how many patients have undergone lower back surgery unnecessarily for what seems to be a disk problem, when the problem was piriformis syndrome all along. Recently, I found myself in a car showroom looking to buy a new car. The showroom manager was clearly in terrible pain. He was hunched over and was extremely reluctant to bear weight on his left leg. I couldn't help myself. I asked him to follow me to his private office, whereupon I proceeded to have him lie on his back on the floor. Keep in mind that he had no clue that I was a neurologist or even a physician for that matter. He followed my instructions while all of the curious sales staff watched through the glass wall of his office.

I asked him to bend his left knee while turning his chin to the left. I gently stretched his piriformis muscle by pushing the bent knee across his body to the right. It was extremely stiff and even slightly painful when I began the piriformis stretch technique, but after a few moments, the spasmodic piriformis began to release. I then had the manager get up and walk around. His pain had completely abated. This was quite a moment at the car dealership.

I can tell you from personal experience that spasm of the piriformis muscle can be incapacitating. It can keep you from working, exercising, and at times can make it difficult to even get out of a chair. Use the exercises on my website to stretch and work the muscle. It's the surest way to keep you moving.

Knee pain is also an extremely common cause of disability. It is the number two cause of chronic pain; more than one-third of Americans report being affected by knee pain. That's more than 100 million people. In the United States alone, more than 600,000 knee replacements are performed annually. By 2030, demand for total knee replacement surgery is expected to exceed $3 million, largely due to older folks staying in the workforce longer and rising obesity rates. There is a time and place for knee replacement, but so often after this procedure is performed, people regret that they consented to it. Surgery should be reserved for the small handful of patients who qualify and who are likely to benefit from it. Most people, however, would do well to avoid it—and all of its risks—and instead focus on strengthening the knees and surrounding muscles.

Lots of people who engage in athletics experience knee pain from something called patellofemoral syndrome. The main sign of this syndrome is pain in front of the knee when sitting, jumping, squatting, or using a staircase—especially going down stairs. Knee buckling, in which the knee suddenly does not support your body weight, is also common. Or you might have a popping or grinding sensation when you are walking or moving your knee. This is usually due to overuse, injury, excess weight, a kneecap that is out of alignment, or structural changes under the kneecap.

So often at the gym, I see people wearing knee braces of various sorts designed to keep the patella in line and alleviate this symptom. Ultimately, though, this tends to make the situation worse. Exercises that can keep the quadriceps and hamstring muscles strong will keep the kneecap where it needs to be, unless there is a significant alignment issue with the legs. Orthotics, corrective inserts in your shoes, may also be helpful.

I myself have had patellofemoral syndrome on both sides, and it is incredible how painful this can be. I once was unable to climb the twenty stairs in my house to get to the bedroom. My orthopedist wanted to inject me with steroids, but I chose to see a physical therapist first, who put me on the path back to health with basic exercises to strengthen my quads. I went from being unable to climb a single staircase to climbing 3,200 feet four months later in just under three-and-a-half hours while traveling in New Zealand.

If you're experiencing pain anywhere in your body, you need to pay attention to that. That is your signal that something is wrong. It may indicate that you're simply overexerting yourself during your exercise—whether cardio or weight-bearing—and not letting the body recover enough between sessions. It may be that you haven't fully stretched for a specific exercise and have strained a muscle, ligament, or tendon. And it could mean that there is an alignment issue that calls for orthotics. If you're engaging in an exercise and it's causing pain, stop, and reassess. Take care of pain when it strikes by modifying your exercise routine as needed, resting sore muscles, and mixing up your routine to make sure you're using different muscles.

When in doubt about the source of pain, get some help from a physical therapist or, even better, a physiatrist. Physiatrists are physicians who treat a wide variety of medical conditions that affect the brain, spinal cord, nerves, bones, joints, ligaments, muscles, and tendons. Go online to learn more about this specialty and find a practitioner in your area.

MAKE ROOM FOR SLEEP

When was the last time you had a good night's sleep? If it wasn't last night, then you're not alone. One in five of us has difficulty sleeping. I've written a lot about sleep in the past, as sleep disorders directly affect the brain, levels of inflammation, and risk for brain issues. The quality and amount of sleep you get have an astonishing impact on virtually every system in your body. Just a generation ago, we didn't think much about the value of sleep other than to refresh the body somehow, like recharging batteries. Today, however, the study of sleep constitutes an entire field of medicine, which has revealed some breathtaking findings about sleep's significance in human health.

Sleep can be described as a "diet of the mind." It repairs and refreshes the brain and body on so many levels; it's no wonder we spend roughly one-third of our life sleeping. Our pituitary gland, for instance, cannot begin to pump out growth hormone until we're asleep. Natural anti-aging growth hormone does more than just stimulate cellular growth and proliferation; it also rejuvenates the immune system and lowers risk factors for heart attack, stroke, and osteoporosis. It even aids our ability to maintain an ideal weight, helping us to burn fat for fuel.

Indeed, getting quality sleep is a requirement for optimum well-being. The better you sleep on a regular basis, the lower your risk for all kinds of health problems. And, conversely, low-quality sleep has far-reaching adverse effects on the body and its functionality. Studies have convincingly shown that our sleep habits impact how much we eat, how fat (or thin) we get, how strong our immune systems are (and whether we can sail through the cold season), how creative and insightful we can be, how well we cope with stress, how fast we can think, and how well we remember things. Prolonged poor sleep habits are a factor in brain fog and memory loss, diabetes and obesity, cardiovascular disease, cancer, depression, and Alzheimer's disease.

In fact, while much has been written about the fact that sleep disturbances are common in patients with Alzheimer's, it was thought that the sleep issues were a consequence of the disease. Newer research, however, indicates that it may be the other way around: Disturbances of sleep may in fact enhance the way the brain makes beta-amyloid protein, a hallmark of Alzheimer's disease. As the authors of a 2015 study state, paying attention to sleep issues and intervening when sleep is not fully restorative may be a way of modifying a risk factor for the future development of the disease.

Here are a few strategies to make the most out of sleep:

> *Prioritize and protect your sleep time.* Just as you would schedule important meetings, schedule your sleep and be brutal about protecting that time period for sleep only. Because the body metabolizes a lot of waste products after 10 p.m. and the immune system revitalizes itself between 11 p.m. and 2 a.m., it's important to be asleep during these hours. So figure out what your bedtime and wake-time should be (e.g., 10 p.m. and 6 a.m.), and don't let anything disturb that pocket of time (see page 171 for more about how to figure out the number of sleep hours you need).

365 days a year. Don't let weekends and holidays derail your sleep routine. Do your best to keep a rigid sleeping schedule every day of the year, no matter what. Your body—and brain—will thank you for it.

Have a sleep study performed on you. The medical term for this study is "polysomnogram." This is a painless, noninvasive procedure that has you spending a night or two in a sleep facility. As you sleep, a sleep technologist records multiple biological functions to determine if you have any disorders like sleep apnea (see next page) or restless leg syndrome. You can find reputable sleep clinics and specialists online.

Watch what you consume. Avoid caffeine late in the day and be mindful of medications you take that might infringe on sleep. Drugs that can undermine sleep include pseudoephedrine (e.g., Sudafed), headache medicines with caffeine (e.g., Excedrin), nicotine, drugs to treat high blood pressure and congestive heart failure, SSRI antidepressants, corticosteroids, and statins.

Cultivate a peaceful, clean sleeping environment. No electronics in the bedroom. Keep it tidy, neat, and at a comfortable temperature for sleeping, which is 65 to 70 degrees Fahrenheit (20°c).

Prepare for sleep. Take time before bed to unwind, disconnect from stimulating activities, and cue the body that it's time for rest. Avoid screens (computers, tablets, and the like) for at least an hour before bedtime. Try a warm bath, listening to soothing music, light reading, or coloring in an adult coloring book. Before lying down, try some deep-breathing exercises (see page 155 for a quick lesson). For some people, physical exercise can bring on restful sleep, but for many, exercising too close to bedtime can be stimulating, preventing you from feeling tired enough to sleep. If you're that type, schedule your regular exercise routine earlier in the day, at least four hours prior to bedtime.

Gear up. Wear loose clothing to bed that's appropriate for the room temperature so you're not too hot or cold.

Try melatonin. If your circadian rhythm is off, which can happen from time to time if you travel across time zones or force your body out of its preferred sleep-wake cycle (maybe you stayed up too late or had a long nap in the afternoon), you might want to try a melatonin supplement. You can also try melatonin if you have unexplained trouble sleeping for many days in a row, as this may be a sign that your body's rhythm is off. Melatonin is our body's natural sleep hormone. But it helps control our twenty-four-hour rhythm as well. Released after the sun sets, it slows body function and lowers blood pressure and core body temperature so that we're prepared to sleep. You can purchase melatonin as an over-the-counter supplement. A proper dosage would be 1 to 3 mg at bedtime.

Rule out sleep apnea. As mentioned above, a sleep study can help you determine if you suffer from this increasingly common disorder that robs millions of restful sleep. Sleep apnea can be more serious than previously thought. It causes the airway to collapse during sleep. Your breathing gets cut off multiple times and your sleep becomes fragmented. Loud snoring and dreamless sleep are often signs of sleep apnea (see box). In 2015, an alarming new study published in *Neurology* found that sleep apnea may be a factor in earlier onset of mild cognitive impairment (MCI) and Alzheimer's disease. MCI often precedes dementia. The researchers of the study found that people with sleep apnea developed mild cognitive impairment nearly ten years sooner than those who didn't suffer from breathing problems during sleep. The time span for developing Alzheimer's also seemed to speed up: Those with sleep apnea developed the disease, on average, five years sooner than the sound sleepers. Researchers theorize that the adverse effects of oxygen restriction on the brain may have something to do with this connection, as well as the fact that sleep drives a slew of physiological events that help the brain "freshen up," do some housecleaning, and clear out proteins that can otherwise gunk up nerve cells.

Signs of sleep apnea:

- frequent fatigue and lack of energy
- excessive daytime sleepiness
- frequent nocturnal urination
- nighttime gasping, choking, or coughing
- irregular breathing during sleep (e.g., snoring)
- morning headaches
- gastroesophageal reflux
- depression

Scientists have documented abnormal brain changes in people who suffer from sleep apnea. The good news is these changes can be reversed through treatment. Studies show that white matter irregularities, for example, can improve immensely when sleep apnea is treated. This is usually achieved through the help of a CPAP device. CPAP stands for "continuous positive airway pressure"; the device, which you wear while sleeping, uses mild air pressure to keep the airways open. The benefits can be felt immediately, and research has shown that within a matter of months, those brain changes go back to normal and there's a vast improvement in cognitive function as well as in mood, alertness, and quality of life. Obesity can also trigger sleep apnea, through the extra weight and fat around the neck. People who lose weight often find relief and no longer need the CPAP machine.

People tend to underestimate the value of sleep. It's arguably more important than what we do during the day. Arianna Huffington wrote an entire book about this subject. I encourage you to read *The Sleep Revolution* if you want to learn more about sleep and how to achieve the best kind. She writes: "It's one of humanity's great unifiers. It binds us to one another, to our ancestors, to our past, and to the future. No matter who we are or where we are in the world and in our lives, we share a common need for sleep."

I should note that your newfound dietary choices will work in sync with your newfound sleep habits. As you clean up your diet and reduce inflammation, you'll be increasing your chances for sound, restful sleep. Check out A. K.'s story of transformation:

As someone who watched a bedridden mother die of Alzheimer's, I have a deep personal interest in preventing this disease in my family. I am always alert for leading-edge information on how to prevent this terrible illness.

Before my diet, I had been consuming lots of processed junk food, including diet colas, crackers and chips, and the daily oatmeal my doctor told me to eat. I was on a dangerous path. Once information about a low-carb, high-fat, gluten-free lifestyle was put before me, I instantly realized this was the information I had been waiting for.

I went to an olive oil store and bought a bottle, started eating grass-fed beef, eliminated grains, switched to green tea, and bought some stevia for (occasional) sweetening. I've also begun to eat more organic greens (daily, in fact).

Previously, I also suffered from arthritis in my joints, especially at night, when the pain would wake me up several times. If a change in sleep pattern is any evidence that this diet is working, then it is worth it to me for that reason only. It has only been six weeks, and the change is already AMAZING!!!

REDUCE STRESS AND FIND CALM IN FOUR SIMPLE WAYS

In my book *Power Up Your Brain: The Neuroscience of Enlightenment*, Dr. Alberto Villoldo and I told the story of how science has come to understand the gift of neurogenesis in humans. Although scientists have long proven neurogenesis in various other animals, it wasn't until the 1990s that the focus turned to humans. In 1998, the journal *Nature Medicine* published a report by Swedish neurologist Peter Eriksson in which he claimed that neural stem cells exist within the brain that are continually replenished and can develop into brain neurons.

And indeed, he was right: We experience brain "stem cell therapy" every minute of our lives. We're not stuck with a finite number of brain cells; conversely, the brain is pliable and can continually make new cells and connections. This is known as neuroplasticity. It explains how stroke victims can learn to speak again.

In September 2014, I was fortunate to serve as conference chair for an international symposium exploring the latest research in brain health. Dr. Michael Merzenich, a professor emeritus neuroscientist at the University of California, San Francisco, and one of the leading pioneers in brain plasticity research, explained that lifestyle factors—some of which might surprise you—can indeed affect the ability of the brain to make new connections.

I've already discussed the ways in which we can positively affect the brain, such as physical exercise, getting restful sleep, following a ketogenic diet, and adding certain nutrients like curcumin and the omega-3 fat DHA. These techniques also have the added effect of reducing the stress our brains and bodies endure daily. Stress will always be a part of our lives; the key is to keep unnecessary stress at bay to preserve and promote those neural connections. And there are other means to positively impact the brain and its connections that have nothing to do with what you eat, how much you exercise, and how well you sleep. When we take a moment to change how we view the world around us, and to act in particular ways that further reduce the stress on our bodies, we are actually changing the physical and functional structure of the brain for the better. To this end, let me outline the four additional ways that can help support this outcome.

- Flex your gratitude muscle
- Maintain strong social networks—mostly offline

- Plan personal downtime
- Get out into nature as much as possible

Flex Your Gratitude Muscle

The science has spoken: The more grateful we feel, the more resilient the brain becomes, physically and even emotionally and spiritually.

Let me explain by way of example how I've incorporated the notion of seeking gratitude in the face of adversity into my life. Several months ago, I received an e-mail that linked to a magazine article about me. The article was anything but positive. The author had published the article—a litany of accusatory and derogatory statements about me—just as he was about to launch a new book, so it was clear he was seeking attention for himself and his book. My first response, coming from my more primitive brain center, was one of anger, outrage, and a strong sense of needing to retaliate.

Over the next several hours, I received more e-mails from concerned friends wondering how I would "respond." I vividly recall being on the phone with my literary agent and publisher, and being asked, "What do you plan to do?" My response to them: "God bless him." Though I was initially angry that someone could have attacked me like this, I realized that I owed the author of this highly disparaging article a great deal of gratitude, as it allowed me the opportunity to truly experience the fact that I do not let others define me. This experience was quite positive, as it did nothing more than strengthen my sense of self.

Gratitude has been studied in the laboratory. In 2015 researchers at Indiana University looked at two groups of people who were being treated for depression and/or anxiety. One group was asked to participate in a gratitude writing exercise, while the other group, which acted as the control, was not. The people in the gratitude writing group spent twenty minutes during the first three sessions of their weekly counseling writing thank-you letters to people in their lives. Three months after the counseling ended, the individuals in both groups underwent a clever experiment using a brain scanner.

The participants were placed in a specific type of functional MRI brain scanner and were gifted various amounts of pretend money by imaginary benefactors. To add realism, the benefactors' names and photos appeared onscreen for the participants to see. The researchers told the participants the following: If they wanted to communicate their gratitude for the money, they could donate all or some of the money to a named third party or to a charity. This may seem like an odd experiment because it's so contrived, but the researchers collected real data by telling the participants that someone would receive real cash minus any amount pledged to another party or charity.

What the researchers found was that, on average, the stronger the thankful feelings a person reported, and the more money a person gave away, the more activity they showed in their brain scans, particularly in areas that are not normally associated with emotions. Which means gratitude is a unique emotion that affects the brain in a unique way. Moreover, the researchers discovered that the gratitude exercise had both short- and long-term effects. Not only did the gratitude writing individuals report feeling more gratefulness two weeks after the exercise, as compared to the controls, but their brain scans showed more activity related to gratitude months later. They were still wired to feel extra thankful.

The takeaway here is that gratitude works mainly because it feeds further cycles of gratitude. It is self-perpetuating. As you practice gratitude, you become more attuned to it, which then allows you to enjoy more of its psychological benefits. In the words of the study's authors: "...you could even think of your brain as having a sort of gratitude 'muscle' that can be exercised and strengthened (not so different from various other qualities that can be cultivated through practice, of course). If this is right, the more of an effort you make to feel gratitude one day, the more the feeling will come to you spontaneously in the future."

One of the easiest ways to practice gratitude is to keep a journal designated for this very exercise. Spend two minutes daily, maybe right before bedtime, writing down a few things for which you feel thankful. These can be little things that occurred during your day or larger experiences or notes of thanks for named people who have had a positive impact on you. Try writing a letter to someone thanking him or her for being in your life, and send it!

Maintain Strong Social Networks—Mostly Offline

One of my favorite quotes is from the 1948 song "Nature Boy" by Nat King Cole. He sings about love being the most important thing one can ever learn—loving others and being loved. I once read that people who work with the dying in hospice care facilities often hear similar questions: Am I loved? And did I love well? These people are at a point in their lives when all the trivial sources of stress are gone and all that's left to ponder is their legacy of love. Love is, after all, everything. I am continually reminded of the power of compassion and love—of the social bonds we maintain, whether they are ones that endure for a long time or are brief but nonetheless impactful. Let me share a real-life story that illustrates what I mean.

Thirty years ago, when I had finally completed my residency training, I was offered a job with an established group of neurologists in Naples, Florida. Soon after I began work, I met Mike McDonnell, an attorney who had an office one floor above us. Mike was a well-known personality in South Florida, and we soon became very close friends. We began sharing evenings together, playing guitar, and singing with other friends. Mike became such an important part of our lives that my future wife and I asked him to perform at our wedding ceremony, and in fact, he and his wife, Nina, joined us on our honeymoon. Mike became the godfather of our daughter Reisha.

Mike turned to me for all things medical, and I in turn relied upon him for his incredible legal acumen. In early February 2016, I was unprepared for the text message that I received from his wife. It simply stated, "Need you, Mike is dying." I rushed to our local hospital and found my friend on a breathing machine with his wife and three of his five children at his bedside. I knew at that moment that I had to assume the role of neurologist and immediately examined him. After reviewing his brain scans, I knew Mike had sustained a massive stroke and had essentially no brain function.

I explained the gravity of the situation to Mike's family and friends. We made arrangements to transfer Mike to the intensive care unit. There, his situation remained fairly stable while on life support. Fortunately, this allowed all of his children to make their way to the hospital and share in Mike's last moments of life.

At 11:14 p.m., Mike left us.

I thought about Mike pretty much the entire next day and later that evening. As fate would have it, one of Mike's close friends, and ours as well, was playing piano at a local restaurant. During his performance, he mentioned how we had lost a close friend the day before. After dinner that night, we spent time with friends talking about Mike and his passing. When we got home, I became violently ill, with shaking chills and nausea. I finally fell asleep about 2 a.m., and when I woke up the next day, I knew something wasn't right.

Plans were made to have a celebration of Mike's life the following day, and we all gathered up our photographs for a retrospective. As it turned out, not only did we have lots of photographs, but we also had a DVD of a performance that our musical group had given at a fund-raiser many years before. After watching the video with my wife and daughter, I felt I needed to lie down on the couch. I wasn't sure what was going on with me, but I was certainly lightheaded and my heart was racing. I then began to lose my vision. I called for my wife and let her know how I was feeling, and she called 911. Prior to the ambulance arriving, fire rescue was in my living room. A young man asked me how I was feeling, and I explained how my heart was racing. He then asked if I was under any stress or had experienced a stressful event, at which point I burst into tears and explained the loss of my friend. The firefighter believed my symptoms were anxiety related, and he encouraged me to take deep breaths and try to relax. My medical mind, while accepting the fact that I was certainly in an anxious state, nevertheless told me that there was something more going on, especially when I took my own pulse and found that it was not only rapid but irregular.

When the ambulance arrived, it was clear that my heart was beating erratically, with a heart rate as high as 170. I was taken to a local hospital, where medicine was administered intravenously to slow my heart, but it failed, twice. At that point, I was transferred to the intensive care unit. The medicine to slow my heart rate was progressively increased, but my heart rate remained dangerously elevated. Finally, Bob, the intensive care nurse, explained that I had reached the maximum dosage of the medication and that a second medication needed to be added. I knew if the medications failed that I would be looking at having a cardioversion procedure done, which is a nice way of saying I'd have my heart shocked back into a normal rhythm.

As evening arrived, I began having a conversation with Bob. He explained that he had worked as a nurse in the trauma unit of an emergency room and described some of his experiences there. As I listened to his stories, I was so taken by his compassion toward me and his wish for me to get better. He carefully adjusted my medications while continuing to tell me about some of the most meaningful events in his life.

As he continued, I closed my eyes and suddenly felt a wave of intense gratitude, not only for my friendship with Mike, but also for my newfound bond with Bob, who was providing care and sharing his life story with me. I can only describe this feeling as one of love. And it was at that moment, as my body was flooded with this emotion, that my heart rhythm suddenly converted back to normal.

As you can imagine, sleeping in an intensive care unit is difficult. I was in and out of sleep throughout the night, and every time I awakened, I checked the cardiac monitor that was behind my bed to make sure that my heart rhythm remained normal. A little after 4 a.m. I woke up again, but this time there was no pulse on the screen. It was flatlined. I thought I was dreaming, but I was awake. I reached down and found that one of my heart monitor leads had become dislodged. I quickly reconnected it, which immediately restored my normal heart tracing to the monitor.

By the time the cardiologist arrived the next morning, I had been out of bed doing yoga. My heart and all my vital signs were deemed normal, and I was discharged with no medication except the recommendation of aspirin.

So many times over the years, I've been asked the question, "What was it that got you into integrative medicine?" I've always said that there was no single epiphany. But I can truly state that the experience in the hospital—first with Mike and then with Bob—has been a major and pivotal event in my life. I left the hospital a changed man. While over the years I have lectured and written about the detrimental effects of stress, those events certainly made a believer out of me. But far more important, they brought me to a place of fully understanding the meaning of love. While we have love for our families and friends, having love and gratitude for others—even strangers—is something that was, at least for me, unexpected but is now fully welcomed in my life. And this was Mike's final gift to me. He actually passed within the same year my father did. My dad was a dedicated physician who throughout his life emphasized to everyone around him the fundamental importance of compassion toward others.

When you are loved and you love well, every cell in your body is allowed to work at its maximum capacity. If love is the single most important ingredient for health and wellness, then I know of no better way of staying on the path of continual healing than to love as much as you can and enjoy its returns. And you do that by maintaining strong social networks. Be open to unexpected newcomers into your life like Bob, and nourish the bonds that have been with you for a long time. You never know when you might need to call on those people when you face serious challenges or have to endure a tragedy in your life.

Social relationships no doubt also change our physiology and sense of well-being. You'd be surprised by how the nature of our health hinges on the nature of our relationships, from the ones we have with others to those we have with ourselves. We are, after all, very social creatures. Recent research has even shown that the bonds we keep with others can enhance longevity. In 2015 a team of researchers at the University of North Carolina at Chapel Hill sought to understand how social relationships affect health. In particular, they were interested in how social relationships "get under the skin" to influence physiological well-being as people age. Among the questions they wanted to answer: When do these effects emerge in life? What do they entail? Do they change as a person ages? How long do they last?

By synthesizing data from four large surveys of Americans ranging from adolescents as young as twelve to seniors as old as eighty-five—a total of more than 14,600 people—the researchers looked at several parameters. In terms of social bonds, they considered social integration, social support, and social strain. To analyze the biological side to the experiment, they considered four common measurements of health: body mass index (a factor of height and weight), waist circumference, blood pressure, and C-reactive protein to assess systemic inflammation. These biomarkers are associated with risk for many diseases, including heart disease, stroke, dementia, and cancer. The results were not surprising on some levels, but they were astonishing on others. We already knew from previous research that older individuals who had a larger social network tended to live longer than those who didn't. But this was the first time a study showed that social bonds lowered health risks in *everyone*—from the young to the old. Among the more surprising findings: social isolation in teenagers contributes equally to bad inflammation as does physical inactivity, and having a strong social network may protect against obesity; social isolation in seniors can be a bigger factor than diabetes in the development and management of high blood pressure; and in middle-aged people, the quality of one's social bonds is more important than the quantity.

There is a lot we can take away from this comprehensive, innovative study, regardless of which stage of life we are in. The relationships we maintain matter to our health, and quality trumps quantity. How well do you relate to other people? Do you have a trusted set of friends? Is your marriage enriching or a source of hardship and stress? Does bad news or a bully in your circle of friends, colleagues, or acquaintances affect you in a way that severely lowers your quality of life? Do you like *yourself*?

Cultivating healthy relationships starts with establishing a healthy relationship with yourself first. This will then allow you to extend that inner love to others and all that surrounds you. And the happier you are in your relationships, the easier it will be to make excellent decisions in all that you do.

Even though there are more gadgets and apps to connect with others than ever, there are also more lonely people who harbor feelings of disconnectedness. It seems as though the more artificial connections we make through social media, the less time we spend with one another in person. To this end, try to nourish your relationships in authentic, intimate ways. Plan more time with the people who inspire, encourage, and de-stress you. And don't rely on social media. Social media platforms have their place in the world, but you can't substitute them for real, face-to-face interactions. Get out there and do things with others. Try new hobbies together. Some ideas:

- Designate a date night (e.g., movie and dinner) with a spouse or best friend at least once a week or twice a month. This doesn't mean you have to go out. Cook the meal together at home and have movie night on the couch.
- Have weekly dinner parties with close friends. Make it a potluck and tell people what to bring.
- Establish a hiking or power-walking group with friends that meets weekly on a designated morning.
- Pick up the phone on weekends and call at least one good friend who lives far away. Catch up.
- Keep a daily ritual with the person closest to you—a spouse, best friend, maybe a child old enough to engage in this exercise. The ritual can be any of a number of things, from

simply talking about your day and what's on your mind to sharing a passage from a book of quotes or proverbs. Each morning, my wife and I share a quotation or passage. For us, it serves to remind us of what is important and meaningful in life. It also helps us to bond. We find that our morning reading stays with us through the day. I often revisit our quote during the course of the day.

- Have a few non-negotiable habits in your life, such as leaving work by 5:30 p.m. so that you can be home to have dinner with the kids. Savor Sundays with family by ditching all electronic devices and focusing on those in-person relationships. Have at least one device-free day a week.

Strengthening those personal connections in real life can be as powerful as any other strategy to support health and well-being.

Plan Personal Downtime

Do you ever try to power through feelings of illness, pain, anger, frustration, exhaustion, being overwhelmed, and the yearning to take a break? Do you have personal downtime on a regular basis? It sounds cliché to say "relax" because it will "reduce stress," but it's all the more important today because we seem to value busy-ness so much. Technologies provide opportunities to be endlessly entertained and engaged, but also distracted and spent. A few neuroscientific studies are starting to emerge, for example, showing that we may be impairing our reflective abilities by relying too much on electronic devices. We use our phones more than we think: In a 2015 paper in the journal *PLOS One*, it was found that people sorely underestimate their smartphone use.

You might think you use your phone an average of thirty-seven times throughout the day, which is what the study's participants believed, but that number is closer to eighty-five! And the total amount of time you spend on your phone daily is probably a little over five hours.

More than a quarter of our days are now spent immersed in information overload. Some of that information is valuable, but some of it is the equivalent of junk food for the brain. The massive digital input could be preventing us from learning and remembering information or from being creative. Some of us don't even take advantage of paid vacation time anymore. But it's essential that we plan personal downtime to let our bodies recover from stress, renew themselves, and gather more strength and energy. Get into the habit of having insightful, distraction-free conversations with yourself during that downtime. Make sure that inner dialogue helps you to stay positive, upbeat, and present.

Scientists at the University of California, San Francisco, have documented that when rats experience something novel, such as encountering a new area, their brains show new patterns of activity. That fresh experience, however, cannot become a solid memory in the rats' brains if they don't take a break from their exploration. The researchers reckon that the findings also apply to how we learn as humans. Downtime allows the brain to take a break and solidify experiences it's had, turning them into permanent long-term memories. If the brain is constantly stimulated, this process could be hampered.

Self-care begins with self-discovery. It's important that we stop to collect our thoughts and check in on our goals regularly. This should be done daily, weekly, monthly, and annually. Some ideas:

- On a daily basis, set a time during which you turn off your cell phone and don't respond to non-emergency calls, e-mails, or text messages so you can practice some deep breathing. This will calm your mind and body and help you evaluate how you feel and what you're thinking. Here's how to do that: Sit comfortably in a chair or on the floor. Close your eyes and make sure your body is relaxed, releasing all tension in your neck, arms, legs, and back. Inhale through your nose for as long as you can, feeling your diaphragm and abdomen rise as your stomach moves outward. Sip in a little more air when you think you've reached the top of your lungs. Slowly exhale, pushing every breath of air from your lungs. Continue for at least five rounds of deep breaths. Then open your eyes and ask yourself whether or not your body feels good and energetic in general. For some, first thing in the morning upon getting out of bed—before looking at any digital device—is an ideal time for deep breathing. Or set the alarm on your phone for 3 p.m. every afternoon. Make it part of your daily routine. Another idea is to end your deep-breathing session with an inspirational quotation. See the box on page 156 for examples.

- On a weekly or monthly basis, tune into broader questions about yourself, such as whether or not you're feeling content, how you're feeling physically, and the status of your relationships. Is there someone you should spend more time with? Anyone you'd be better off excluding from your life? What in life is causing you a great deal of stress and anxiety? How can you begin to remedy that?

- On an annual basis, set new goals and address any challenges. Consider the big goals, such as anything you might want to accomplish that may require long-term planning. What would you like to do in the next year or the next decade? Find a new job?

Perfect a skill? Try a new hobby? Start a business? Climb Mount Kilimanjaro? Travel around Europe? Volunteer more? Enrol in an art workshop? Go on a weekend retreat? Write a memoir?

As mentioned earlier, recording your thoughts, goals, feelings, anxieties, and the events most affecting you in a journal can be helpful. It allows you to review later, can help assuage those worries, and gives you accountability (for a list of the journals to maintain, see page 172).

Reading a short but meaningful quotation is a great way to cap a deep-breathing session. Here are thirty suggestions to get you started:

1. If you do not change direction, you may end up where you are heading. — Lao Tzu
2. If not now, when? — Rabbi Hillel
3. The best and most beautiful things in the world cannot be seen or even touched — they must be felt with the heart. — Helen Keller
4. We must let go of the life we have planned, so as to accept the one that is waiting for us. — Joseph Campbell
5. Our greatest weakness lies in giving up. The most certain way to succeed is always to try just one more time. — Thomas A. Edison
6. Consult not your fears but your hopes and your dreams. Think not about your frustrations, but about your unfulfilled potential. Concern yourself not with what you tried and failed in, but with what it is still possible for you to do. — Pope John XXIII
7. If you want others to be happy, practice compassion. If you want to be happy, practice compassion. — Dalai Lama
8. Perseverance is not a long race; it is many short races one after the other. — Walter Elliot

9. Patience and perseverance have a magical effect before which difficulties disappear and obstacles vanish.—John Quincy Adams
10. As we express our gratitude, we must never forget that the highest appreciation is not to utter words, but to live by them.—John F. Kennedy
11. True success is overcoming the fear of being unsuccessful.—Paul Sweeney
12. Peace is not absence of conflict; it is the ability to handle conflict by peaceful means.—Ronald Reagan
13. Once we accept our limits, we go beyond them.—Albert Einstein
14. God grant me the serenity to accept the things I cannot change, the courage to change the things I can, and the wisdom to know the difference.—The Serenity Prayer
15. The only real security is not in owning or possessing, not in demanding or expecting, not in hoping, even. Security in a relationship lies neither in looking back to what it was, nor forward to what it might be, but living in the present and accepting it as it is now.—Anne Morrow Lindbergh
16. Never give up, and be confident in what you do. There may be tough times, but the difficulties which you face will make you more determined to achieve your objectives and to win against all the odds.—Marta Vieira da Silva
17. The friend in my adversity I shall always cherish most. I can better trust those who helped to relieve the gloom of my dark hours than those who are so ready to enjoy with me the sunshine of my prosperity.—Ulysses S. Grant
18. Faith is the bird that feels the light when the dawn is still dark.—Rabindranath Tagore
19. There are two great days in a person's life—the day we are born and the day we discover why.—William Barclay

20. There is no end to education. It is not that you read a book, pass an examination, and finish with education. The whole of life, from the moment you are born to the moment you die, is a process of learning.—Jiddu Krishnamurti
21. Your task is not to seek for love, but merely to seek and find all the barriers within yourself that you have built against it.—Jalal Al-Din Rumi
22. Let the one among you who is without sin be the first to cast a stone.—Jesus Christ
23. You can search throughout the entire universe for someone who is more deserving of your love and affection than you are yourself, and that person is not to be found anywhere. You yourself, as much as anybody in the entire universe, deserve your love and affection.—Buddha
24. We ourselves feel that what we are doing is just a drop in the ocean. But the ocean would be less because of that missing drop.—Mother Teresa
25. We must welcome the future, remembering that soon it will be the past; and we must respect the past, remembering that it was once all that was humanly possible.—George Santayana
26. Despite everything, I believe that people are really good at heart.—Anne Frank
27. The only thing worse than being blind is having sight but no vision.—Helen Keller
28. In the end, it's not the years in your life that count. It's the life in your years.—Abraham Lincoln
29. It does not matter how slowly you go as long as you do not stop.—Confucius
30. Accept the challenges so that you can feel the exhilaration of victory.—George S. Patton

Get Out into Nature as Much as Possible

Our ancestors used to work and live mostly in the outdoors, but few of us do that anymore. We live and work indoors, usually tethered to electronics, chairs, couches, meetings, and chores. There is a biological reason why going for walks and hikes, for example, or doing anything in the open air can be so invigorating. Being outside and among plants and other living things enhances feelings of well-being through a variety of biochemical reactions, including a real calming effect on your mind and nervous system.

Get out in nature as much as possible during the day, whether you live in a big city or a rural area. Find a park to take a daily walk after lunch. Try to sit near a window with a view when at work, or place a chair in front of the window with the best view from your home. Notice the movement of trees in the wind, the nearby birds and other creatures. Plan your workouts outside when the weather is agreeable. Take in the air and scenery along natural bodies of water or mountainous regions. Enjoy the first light of the day at dawn and the sunsets at night. On clear nights, go stargazing. And don't forget to bring the outdoors inside. Decorate your rooms and office with living plants (see the next section for ideas). They will keep your air clean and bring Mother Nature closer to you. And as you're about to read, they will help you to detoxify your physical environment.

DETOXIFY YOUR PHYSICAL ENVIRONMENT

Let me state the obvious. We live in a sea of chemicals. Scientists who measure the so-called body burden, or levels of toxicants in tissues of the human body, tell us that virtually every resident in the United

States, regardless of location or age, harbors measurable levels of synthetic chemicals, many of which are fat-soluble and therefore stored in fatty tissue indefinitely. I wish there were more focus on policing these chemicals than on monitoring them. Unfortunately, it takes years—sometimes decades—for studies to gather enough evidence for the government to justify writing new standards or regulations, and even to take dangerous products off the market. In 2014, a meta-analysis published in the *Journal of Hazardous Materials* reviewed 143,000 peer-reviewed papers to track the patterns of emergence and decline of toxic chemicals. The study exposed the sad truth: It takes an average of fourteen years for appropriate action to take place. We need to take matters into our own hands.

The good news is the Grain Brain Whole Life Plan helps you do just that. Don't wait until something is officially labeled as being dangerous to eliminate it from your life; when in doubt, take it out of your life.

I've already made my case against glyphosate, the main ingredient in Roundup. Here are some additional ideas to support a cleaner way of life:

- When buying canned foods, be sure the cans are not lined with BPA. Look for "BPA-free lining" language on the can.
- Avoid using nonstick pans and other cookware. Teflon-coated wares contain perfluorooctanoic acid, or PFOA, which the EPA has labeled a likely carcinogen. Cast-iron cookware, ceramic, uncoated stainless steel, or glass are your best bet.
- Minimize the use of the microwave. Never place plastic—including plastic wrap—in a microwave. Don't put hot foods in plastic, which can release nasty chemicals that are absorbed by the food.

- Avoid plastic water bottles, or at least avoid plastics marked with "PC," for polycarbonate, or the recycling labels 3, 6, or 7 on the little triangle. Buy reusable bottles made of food-grade stainless steel or glass.

- When it comes to toiletries, deodorants, soaps, cosmetics, and general beauty products, switch brands when you restock. Remember, your skin is a major entry point to your body, and what you slather on may make its way inside to inflict harm. Look for organic certification and choose products that are safer alternatives. Use the Environmental Working Group's (EWG) user-friendly website (www.ewg.org) and I Read Labels For You (www.ireadlabelsforyou.com) to find the safest products. Endocrine-disrupting chemicals, or EDCs, have been shown to disrupt normal metabolism and even trigger weight gain. The most insidious ones are:
    - aluminium chlorohydrate (in deodorants)
    - diethyl phthalate (in perfumes, lotions, and other personal care products)
    - formaldehyde and formalin (in nail products)
    - "fragrance" and "parfum" (in perfumes, lotions, and other personal care products)
    - parabens [methyl-, propyl-, isopropyl-, butyl-, and isobutyl-] (in cosmetics, lotions, and other personal care products)
    - PEG/ceteareth/polyethylene glycol (in skin care products)

- sodium lauryl sulfate (SLS), sodium laureth sulfate (SLES), and ammonium lauryl sulfate (ALS) (in a variety of products: shampoos, body washes and cleansers, liquid hand soaps, laundry detergents, hair color and bleaching agents, toothpastes, makeup foundations, and bath oils/bath salts)
- TEA (triethanolamine) (in skin care products)
- toluene and dibutyl phthalate (DBP) (in nail polishes)
- triclosan and triclocarban (in antibacterial hand soaps and some toothpastes)

- Select household cleaners, detergents, disinfectants, bleaches, stain removers, and so on that are free of synthetic chemicals (look for brands that use natural, nontoxic ingredients; again the www.ewg.org site can be helpful here). Or make your own: Simple, inexpensive, and effective cleaning products can be made from borax, baking soda, vinegar, and water (see the box on page 163).

- Indoor air is notoriously more toxic than outdoor air due to all the particulate matter that comes from furniture, electronics, and household goods. Ventilate your home well and install HEPA air filters if possible. Change your air-conditioning and heating filters every three to six months. Get the ducts cleaned yearly. Avoid air deodorizers and plug-in room fresheners. Reduce toxic dust and residues on surfaces by using a vacuum cleaner with a HEPA filter. Naturally ventilate your house by opening the windows.

- Request that people take off their shoes upon entry.
- Plants—such as spider plants, aloe vera, chrysanthemums, gerbera daisies, Boston ferns, English ivies, and philodendrons—naturally detoxify the environment. Keep as many in your home as possible.
- When purchasing clothes, fabrics, upholstered furniture, or mattresses, choose items that are made of natural fabrics with no flame-retardant, stain-resistant, or water-resistant coatings. (Some states mandate a certain level of flame retardants on products, but do your best to find the most natural products possible.)
- Wet-mop floors and wipe down windowsills weekly.
- Speak with your local garden store or nursery personnel for recommendations on pesticide- and herbicide-free products you can use in your garden to control pests.

THREE HOMEMADE CLEANING PRODUCTS

All-purpose cleaner and deodorizer:

115g bicarbonate of soda

2 liters warm water

Combine the ingredients and store in a spray bottle.

Glass and window cleaner:

1 liter water

250ml white vinegar

125ml 70% rubbing alcohol

2 to 4 drops essential oil (optional, for aroma)

Combine the ingredients and store in a spray bottle.

Disinfectant:

2 teaspoons borax

4 tablespoons white vinegar

750ml hot water

Combine the ingredients and store in a spray bottle.

While it may seem like an overwhelming task to clear out your house of questionable products and replace them with alternatives, it needn't be stressful and you needn't do it all in one day. Go one room or one product at a time. The goal is to do the best you can based on what you can afford and what you're willing to change. As part of your daily checklist during the 14-day menu plan, I'll ask you to do one thing that helps you to detoxify your physical environment.

But before we get to that, there's one more step to take that will help you pull all these ideas together: Plan accordingly.

CHAPTER 7

Step 3—Plan Accordingly

THE BODY LOVES AND CRAVES consistency and predictability. So much so, in fact, that it will rebel in subtle ways if you force it to be out of sync with its natural rhythms. This is partly why traveling across multiple time zones and temporarily abandoning your usual routine can feel so difficult and uncomfortable. Your smart body will do everything it can to get you back on track quickly. One of the easiest ways of reducing unnecessary stress on your body and maintaining a balanced, homeostatic state is to keep a steady daily routine year-round to the best of your ability, including weekends and holidays. This demands that you plan your days well in advance.

Work obligations, social demands, and unexpected events have us all breaking this rule occasionally, but see if you can regulate at least the three aspects of your life that will have the biggest impact on your health: when you eat, when you exercise, and when you sleep. If you do, you will notice a difference in how you feel. It may also help you to stick to a routine and be better prepared for unforeseen challenges that can derail your new way of life under the Grain Brain Whole Life Plan.

So to that end, let me give you some guidance.

WHEN TO EAT

Calories cannot tell time, but the body can, and it will receive calories differently depending on a variety of factors, including—you guessed it—the time of day.

Each us has an internal system of biological clocks that help the body manage and control its circadian rhythm—your body's sense of day and night. This rhythm is defined by the patterns of repeated activity that correlate with the twenty-four-hour solar day and include your sleep-wake cycle, the rise and fall of hormones, and changes in body temperature. Just recently we've discovered that when these clocks are not "on time"—when they're not functioning properly—disordered eating and weight problems can result. It's well documented, for example, that obese people often have disrupted circadian rhythms, triggering them to eat frequently and at irregular times, especially late at night. Obese people also often suffer from sleep apnea, which further disturbs their sleep rhythm. And, as you already know, sleep deprivation can impact the balance of those appetite hormones leptin and ghrelin, further exacerbating problems.

In Chapter 4, I gave you some parameters for intermittent fasting, suggesting that you skip breakfast once or twice a week and fast for seventy-two hours four times throughout the year. As an additional tip, I recommend that you *eat more of your daily sum of calories before 3 p.m.* and avoid eating a bounty of food at night. In fact, *avoid eating anything within four hours of bedtime.* (You can drink water and caffeine-free tea, but try not to drink anything within a half an hour of bedtime. Otherwise, you might have to get up in the middle of the night to use the bathroom.)

The power of eating lunch before 3 p.m., for example, was recently highlighted by researchers at Harvard's Brigham and Women's Hospital, Tufts University, and Spain's University of Murcia, who conducted their study in the Spanish seaside town of Murcia. Spaniards make lunch their main meal of the day. To the researchers' surprise, all things being equal, such as total calories eaten daily, levels of activity, and sleep quantity, those who ate lunch later in the day struggled more with weight loss. All of the participants—a total of 420 individuals—in the study were either overweight or obese. And all of them were put on the same five-month weight-loss program. But they didn't eat lunch at the same time, and they didn't experience the same weight loss. Half of them ate lunch before 3 p.m., and the other half ate after 3 p.m. Over the course of the twenty weeks, the early lunchers lost an average of twenty-two pounds (10kg), while the later lunchers shed only seventeen (7.5kg) and at a slower clip.

We intuitively know that overeating toward the end of the day is not a good idea. That's when we're likely to be tired. Even our brains are tired of making decisions, so we cave in to a mindless feast at the dinner table with multiple portions of unhealthy food choices and dessert. This is especially the case when we've had a busy day, skipped lunch entirely, and grazed on nutrient-poor snacks and food products. And if it has been a very long time since we've eaten, and we haven't taken in enough calories earlier in the day, it's far too easy to gorge at dinner because the body will want to make up for those lost calories. It's a biological and metabolic assault. I recommend the following to avoid this scenario and maximize your body's energy needs:

- Over the weekend, plan your upcoming week's eating schedule and meals based on what's on your calendar in terms of work and personal responsibilities. Use a journal for this planning (see page 172). Choose the one or two days you'll skip breakfast. On those days, be sure to have a nutrient-dense lunch between 11:30 a.m. and 1:30 p.m.

- Create your grocery list based on the meals you plan to prepare. Don't forget to include snacks. Take care of that shopping before the week begins. You don't want to wake up Monday morning and be scrambling to find food to eat for breakfast or to pack for a lunch.

- Identify which meals and snacks you can take with you for lunch when you're away from home.

The more you plan your eating habits, from *when* you eat to *what* you eat, the more you can have effortless control over sticking with the plan and reaping its health rewards. The same is true when it comes to exercise.

WHEN TO EXERCISE

The body might be physically stronger in the late afternoon and early evening due to a peak in body temperature and certain hormones like testosterone, but that doesn't mean you have to engage in physical activity at that time. You should schedule your exercise when it works best for you. It's far more important to do the exercise than to worry about the "best" time of day for your body to be active. Some people enjoy an early morning jog, while others prefer to end their day with exercise.

Keep in mind, though, that an hour daily spent working out hard won't erase the effects of sitting down for the rest of the day. A growing body of research is revealing that it's entirely possible to get plenty of physical activity and still suffer increased risk of disease and death—just like smoking harms no matter how much you engage in other, healthier lifestyle habits. So many of us barely move an inch to go from home to car to office chair, and back again to sit on the couch and watch TV. It behooves all of us to work more movement into our day, no matter what kind of job we have. Get creative with turning to-do's that could be done without much effort into tasks that require physical movement. Take the stairs. Park far away from the building you plan to enter. Use a headset to walk and talk on your phone so you're getting up from your desk. Take a twenty-minute walk during your lunch break. Build more opportunities to be active during your day. Some additional tips:

- When you map out your upcoming meals for the week, plan your exercise time as well. Also plan what kind of exercise you will do, using a journal for this planning (see page 172). Remember, the minimum is twenty minutes of cardio six days a week with weight-bearing exercises three or four times a week. Build in extra time for stretching, too. Decide which days you'll engage in vigorous exercise and which will be less intense (see the sample plan on the following page). On that seventh day of "rest" (which doesn't need to be a Sunday), plan to do something low-key, like going for a walk with a friend or taking a meditative yoga class. A day of rest doesn't mean you stay totally inactive on your derriere all day.

- If you suffer from sleep disorders or have a hard time falling asleep at night, try to break a sweat outdoors in the early morning hours. Exposure to daylight (say, during a morning bike ride, jog, or drive to an exercise class) soon after waking effectively reboots your circadian rhythm. Morning exercise has also been shown to reduce blood pressure throughout the day and cause an additional 25 percent dip at night, which further correlates with better sleep.

Here is a sample exercise plan for someone who already has a baseline level of fitness and is hoping to gain more strength and fitness with higher-intensity workouts and longer stretches of moderate activity throughout the week. Note that Sunday doesn't have to be the "off" day of rest—here, it's Wednesday. Plan those longer workouts on days when you have more time, which for many is over the weekend.

Monday: Midday brisk walk (twenty to thirty minutes); weight training and stretching at the gym after work (twenty minutes)
Tuesday: fifty-minute indoor cycling class in the morning, plus ten minutes of stretching
Wednesday: Crazy busy day – thirty minutes of brisk walking anytime during the day, and fifteen minutes of weight-bearing exercises and light stretching while dinner is cooking
Thursday: Elliptical machine (thirty minutes) in the morning, plus ten minutes of stretching
Friday: Vinyasa flow yoga class at 6 p.m.
Saturday: Weekend Warrior Power Walking Group at 9:30 a.m. (ninety minutes)
Sunday: Elliptical machine (forty minutes), plus weight-bearing exercises and stretching (twenty minutes)

The more specific you are with your formal exercise plan during the week, the more likely you are to stick to it.

WHEN TO SLEEP

Remember, the body—and especially the brain—revitalizes itself during sleep. While we used to think that there was a magic number of hours the body needed to sleep, new science has overturned that myth. Everyone has different sleep needs. How much do *you* require to function optimally? Find out:

- Determine an ideal wake-up time given your morning duties
- Set your morning alarm for that time every day
- Go to bed eight to nine hours prior to that time until you wake up *before* your alarm. The number of hours of sleep you get that night is your ideal number

Some additional reminders:

- Use the strategies I outlined in Chapter 6 to prepare for sleep and make the most of it
- Be strict about going to bed and getting up at the same time every day, 365 days a year. Don't shift your sleep habits on weekends or holidays or while on vacation

Aim to be asleep before 11 p.m. The hours between 11 p.m. and 2 a.m. are critical for health. This is when your body's powers of rejuvenation are at their peak.

A DAY IN THE LIFE

I've been mentioning the use of journals throughout the book. I can't think of a better way to plan and track your daily life than to keep a few journals for various purposes. This automatically holds you accountable to your intentions and goals. Here's a summary of the three main journals I recommend:

- Food journal: This is where you keep a running tab of not only what you eat, but also what you *plan* to eat—your meals and snacks throughout the week. On the weekend, look ahead at the upcoming week and map out your daily menu, then write down exactly what you eat each day and see how close you come to living up to your plans. Note which foods and ingredients you like or dislike, writing down your favorite meals and recipes. Add details like which foods make you feel extra good and which others may be problematic for your unique physiology. If you tweak a recipe to your liking, record that.
- Exercise journal: This is where you maintain your plans for exercise and record what you do in fact accomplish each day. Track your minutes of cardio, strength training, and stretching. List which muscle groups you work and what type of cardio workout you do. If you experience any pain or soreness, jot that down. See if you can find patterns in the exercises you choose and how your body feels, as this can help you tailor the right exercise regimen for your body. Over the course of a week, make sure you're mixing up your routines, going hard on some days and lighter on others.
- General journal: This is where you document your thoughts, feelings, ideas, wishes, goals, and notes of gratitude. Don't hesitate to record your worries and anxieties, as writing those down can have the effect of reducing their psychological impact on you.

It doesn't matter what type of books you use for journaling. You can buy inexpensive spiral-bound notebooks for your food and exercise journals and splurge on a leather-bound diary for your general journal. Keep a journal by your bedside for early morning and evening writing, and take small, convenient notebooks with you wherever you go to jot down notes throughout the day. Do what works for you and keeps you on track.

Below is a daily checklist followed by a sample daily schedule.

Your Daily Checklist

- ❑ Get up and go to bed at the same time daily.
- ❑ Take your supplements, including your prebiotics and probiotics. See page 112 for your cheat sheet about which supplements to take, how much, and when.
- ❑ Unless you're skipping breakfast, which I encourage you to do at least once a week, make sure you're getting a little protein in the morning. Remember that eggs are a perfect way to start the day.
- ❑ Do cardio exercise for a minimum of twenty minutes, with stretching before and after. Every other day, do weight-bearing exercises (see www.DrPerlmutter.com for videos). See page 168 for information about timing your exercise.
- ❑ Do one small thing to clean up your physical environment (see page 159).
- ❑ Eat lunch before 3 p.m.
- ❑ Drink water throughout the day.
- ❑ Take a ten-minute distraction-free timeout in the a.m. and p.m. to check in with yourself, maybe do some deep breathing (see page 155), write in a journal, or read an inspiring quote or

passage from a book. If you'd like to try meditation, go to www.how-to-meditate.org/breathing-meditations.

- ❑ Plan dinner so that it's not within four hours of bedtime.
- ❑ Try to be in bed with the lights out before 11 p.m.

Sample Daily Schedule

6:30 a.m.	Wake up!
6:30–6:45 a.m.	Morning deep-breathing exercise and journal writing
7:00–7:45 a.m.	Exercise (e.g., stationary bike, weight training, and stretching)
7:45–8:15 a.m.	Bathing and grooming
8:15 a.m.	Prepare breakfast and bagged lunch
8:45 a.m.	Out the door for work
12:30 p.m.	Lunch and 20-minute walk
4:00–4:15 p.m.	Snack and a few minutes of self-reflection
5:30 p.m.	Leave work
6:30 p.m.	Supper with kids
7:30–8:00 p.m.	Personal downtime
9:30 p.m.	Cutoff time for electronics, prepare for bed
10:30 p.m.	Lights out!

While there are plenty of apps out there to help you map out your day and send reminders to your phone through texts, there is nothing wrong with using an old-fashioned daily planner. Do what works for you. Get as detailed as you like, but understand that everything in your life should revolve around your eating, exercising, and sleeping patterns. Be consistent, even selfish, with those routines, and your whole body will reap tremendous health benefits. I hate to be cliché, but it's true: Timing is everything.

CHAPTER 8

Troubleshooting

EVERY MINUTE OF EVERY DAY we get to choose. Like I always say, life is an endless series of choices. Right or left? Yes or no? Fish or french fries? The whole point of this book has been to help you learn to make better decisions that will ultimately allow you to participate in life at its fullest. Even though you will be faced with difficult decisions, I know you can do this. You know the value of being healthy and mentally sharp. You know what sudden illness and chronic disease can do. Your health should be the most important thing in your life. Because what would you do without it?

Christopher E. posted the following story to my website:

I didn't start out unhealthy, but I had let myself get run down through a combination of work stress, physical stress, and poor nutrition. I'm not an old guy and I've always been able to get away with whatever I wanted to do, so why not work eighty hours a week, train to climb a mountain, and knock out the Bataan Death March all in a six-month period, all while getting four hours of sleep a night and supplementing with lots of coffee! Shockingly, I started feeling super-fatigued every day a few weeks after the mountain climb, and then, to my surprise, my hair started falling out. Seriously, patches of hair started falling out! Being an Army officer, I keep my hair short, but one day one of my sergeants said, "Sir, what's going on with the back of your head?" I checked it out and sure enough, I had a bald patch.

Over the months, it kept getting worse, and the dermatologist said it was alopecia and that we might be able to treat the patches with steroid injections. Without that, Doc said they may get better or worse, but he also said the best thing I could do was limit stress. Yeah, right. I got orders to change duty stations, and my wife told me that we were pregnant all in the same month. The patches kept getting worse, and being a leader in the military who also happens to work in health care, it certainly didn't help my leadership presence or put patients at ease to see my patchy head.

Fast forward eight months or so after moving, and I picked up a copy of Grain Brain. *Intrigued by the assertion that some nutritional supplements (gasp) were being touted by a neurologist as being neuroprotective and restorative, along with the idea that the gut microbiome could impact not only the brain but also just about every system in your body, I quickly read the book.*

Feeling and looking terrible, I went ahead and gave the ketogenic diet a shot and I began six of the seven supplements suggested in the book (minus the resveratrol). We had the baby the next month (September 2015), so my stress level and sleep got worse, but the patches magically started filling in. By January they were all gone, and I no longer feel like I'm lifting a piano every time I get out of bed. I've also lost twenty pounds.

The success has inspired and empowered me, and now I'm tackling my stress and sleep. I recently began a daily meditation period (not sure if it's doing anything but I'm hopeful), writing a gratitude list, and I'm trying to get seven hours of sleep every night.

I seriously used to think this approach to health was laughable (I was actually taught that it was garbage), but after my own personal case study, I'm starting to come around.

There will no doubt be challenges as you proceed. And there will be times when you will have to address these challenges one by one. That's life. Following are some troubleshooting tips for those moments that threaten to derail you. By all means this is not an exhaustive list, but it will help you handle those inevitable times when you have to make hard choices.

"HOW DO I FOLLOW THE PLAN WHILE EATING OUT OR AWAY FROM HOME?"

I recommend that you avoid eating out during the first couple of weeks on the program so that you can focus on getting the dietary protocol down using my 14-day meal plan. This will prepare you for the day you do venture away from your kitchen and have to make good decisions about what to order from someone else's kitchen.

Most of us eat out several times a week, especially while we're at work. It's virtually impossible to plan and prepare every single meal and snack we consume, so you'll need to learn to navigate other menus. See if you can order from the menu at your favorite restaurants. Don't be shy about asking for substitutions (e.g., a side of more steamed veggies instead of potatoes; extra-virgin olive oil instead of their commercially produced vinaigrette). If you find it too challenging, then you may want to try new restaurants that can cater to your needs. It's not that hard to make any menu work as long as you're savvy about your decisions. Look for healthy sources of organic, GMO-free vegetables. Then add some fat—a drizzle of olive oil or half an avocado—and a little bit of protein and you'll be fine. Watch out for elaborate dishes that contain multiple ingredients. When in doubt, ask the waiter or chef about the dishes.

Rather than eating lunch out while you're at work, consider packing meals. Having precooked foods—such as roasted or grilled chicken, hard-boiled eggs, poached salmon, or strips of grilled sirloin steak or roast beef—in your refrigerator ready to go is helpful. Fill a container with salad greens and chopped raw veggies and add your protein and dressing of choice before eating. I travel with avocados and cans of sockeye salmon. Canned foods can be excellent sources of good, portable nutrition, as long as you're careful about which products you buy and that the cans are labeled BPA-free.

Keep snacks on hand, too, especially at the beginning of this new way of life when you're cutting carbs. There are plenty of snack and "on-the-go" ideas listed in Part III, many of which are portable and nonperishable.

And when you're faced with temptation (the box of muffins at work or a friend's birthday cake), remind yourself that you'll pay for the indulgence somehow. Be willing to accept those consequences if you cannot say no. But keep in mind that a grain-brain-free way of life is, in my humble opinion, the most fulfilling and gratifying way of life there is. Enjoy it.

"OH, NO! I'VE STRAYED FROM THE PROTOCOL. NOW WHAT?"

Maari C.'s story says a lot about what can happen when you revert to your old ways of eating, or suddenly reintroduce wheat after evicting it from your diet. Even though Maari intentionally strayed from the protocol for a couple of days, the effects were so massive that it's worth learning from her experience:

> *Several years ago, I started suffering from panic attacks and anxiety, and I quickly began researching holistic alternatives to the pharmacy medications doctors wanted to put me on. After three months of these attacks, my health was repaired within a month.*
>
> *Fast forward seven years, and after my second pregnancy I started breaking out in hives all over my body, on top of having an underactive thyroid (a big surprise, as I had an overactive thyroid). I was forced to take Synthroid.*
>
> *A few months ago, my skin rashes came back. On top of that, I was exhausted and depressed. After stumbling across some literature, I adopted a wheat-free diet. As a result, I feel amazing. In two weeks of stopping wheat, my skin cleared up from a rash I've had for seven years! My energy soared through the roof. I didn't get hungry or cranky (amazing, since I was eating 1,200 calories a day on this cleanse). I found myself so excited to drink my green juice for lunch because it made me feel like a million bucks!*
>
> *The other thing that amazed me was that I was no longer dizzy. I was two weeks into "wheat-free" while on the swings with my daughter, and for the first time I wasn't dizzy or nauseous on the swings. It has been amazing!*
>
> *I have been amazed at how my brain feels after this. It feels clearer and I'm more articulate. I did go back on wheat for two days to see how it affected me, however, and within thirty minutes of eating a slice of bread, I felt lightheaded. Later, I had some pizza and was miserable that night. Within the day, I had a cold sore and an eye infection, things I haven't had since college.*

As with so many things in life, discovering and establishing a new habit is a balancing act. Even once you've shifted your eating and exercise behaviors and changed the way you buy, cook, and order food, you'll still have moments when old habits emerge. (And of course I don't endorse intentionally experimenting with old habits again as Maari did. That will derail you physically and emotionally.) I don't expect you to never eat a slice of crusty pizza or drink a beer again, but I do hope that you stay mindful of your body's true needs and live according to these principles as best you can.

Aim to stick to the 90-10 rule: Follow these guidelines 90 percent of the time, leaving 10 percent wiggle room. There is always an excuse for not taking better care of yourself. We have parties and weddings to attend. We have work to address that leaves us high on stress and low on energy, time, and the mental bandwidth to make good food, exercise, and sleep choices. Hit reboot whenever you feel like you've fallen too far off the wagon. You can do this by fasting for a day and recommitting to the protocol. Take a Friday or Monday off from work to enjoy a three-day weekend during which you get out of town to focus on yourself or enjoy a staycation. Maybe try a yoga retreat or visit a friend you haven't seen in a long time. The point is to shake things up a bit, get out of that health-depleting rut, and refresh your resolve to succeed.

"I'M HAVING SERIOUS HEADACHES AND OTHER SIDE EFFECTS FROM THE DIET"

No sooner do you begin this dietary protocol than your body goes into high gear, detoxifying itself and shedding excess poundage. Headaches can be a common response to a sudden shift in your diet, especially if you've been eating poorly. But they're actually a sign that the diet is working, and they will go away within days. If you feel the need to use an over-the-counter pain reliever, try aspirin. Unlike other NSAIDs, which can disrupt the microbiome and blunt emotions, aspirin can relieve pain and also have some anti-inflammatory effects. As noted in Chapter 4, you might also crave carbs and feel a bit moody and irritable at first when you go cold turkey on the carbs. Remember, gluten and sugar can act much like drugs, leaving those quitting them going through a period of withdrawal. This is normal as your body adjusts and goes through the transition away from the processed, packaged (or otherwise low-quality) foods you may have been eating. Your mood will adjust as you go along, but how can you handle the cravings? What can you do to overcome them?

Be assured that cravings will not last long. Many readers of *Grain Brain* have told me that once they went cold turkey and dove right into my protocol, they never again experienced a craving like they had back when carbs were a staple in their diet. It takes some willpower for the first week and then it gets easier. As one of my fans wrote to me, "It is like feeding the dog from the table—if you keep eating even a little bit of the bad stuff your system will keep begging for it."

But if you do experience a gravitational force toward a bread basket, a chocolate chip cookie, or a bowl of steaming hot pasta, try to distract yourself by engaging in a new activity. Shift gears. Go outside for a 20-minute walk. If it's not too close to bedtime, do some formal exercise (e.g., download an exercise video you can follow at home). Take fifteen minutes to write in a journal or perform some deep-breathing exercises. Listen to uplifting music. Tackle a project you've been meaning to do, such as cleaning out a desk drawer or closet. Or simply find something else to eat. Have a snack on hand to settle those cravings: a handful of nuts, or half an avocado drizzled with some olive oil and balsamic vinegar. Try to keep a meal prepared and ready to go in the fridge all the time, so that if you're super hungry or out of energy you don't go for take-out. Remind yourself that the carbs and sugary foods are just filler foods—made to fill you up with inflammation and pain in the long run. Tell yourself that you'd rather fill up with good-quality foods that do your body and mind good. Remind yourself that you are worth it.

"HELP, I'M GOING TO BED HUNGRY"

If you're having a hard time blocking out that four-hour time period after dinner during which you don't eat anything, here's what you can do. Make sure you're getting enough satiating fat at dinner. Add more olive oil to your vegetables or have a small portion (8 tablespoons) of a rich non-gluten grain such as quinoa with a drizzle of olive oil at dinner.

Rather than heading to the refrigerator when you feel hungry toward bedtime, distract yourself. Try drinking some chamomile tea or other warm herbal tea while reading a good book or magazine article. Call a friend (see page 195 about finding a partner for your journey). Take an evening stroll around the neighborhood. Write in one of your journals. If you have young children, play with them or read books to them. The goal is to distract yourself from thinking about food. If you find yourself lying in bed unable to go to sleep, focus on your breathing and keep your thoughts fixated on the health benefits that are happening at that moment.

"I'M VEGAN. WHAT SHOULD I DO?"

A vegan diet can be wonderfully healthy as long as you're getting good sources of vitamins D and B_{12}, and the omega-3 DHA, as well as minerals like zinc, copper, and magnesium. DHA is available as a supplement derived from marine algae, a vegetarian source. Although people sometimes worry that vegans don't get enough protein, they can get plenty from vegetables, legumes, and non-gluten grains. What I worry about most with vegans is that they don't get enough fat due to the exclusion of all animal products, including eggs and fish. So added olive oil and coconut oil will help bring this dietary choice into balance.

ATTENTION PREGNANT WOMEN AND NEW MOMS

You can indeed build a better baby through the strategies outlined in this book. When expectant and new moms ask me for advice, I offer four important tips:

1. Take prenatal vitamins and probiotics.

2. Supplement with 900–1,000 mg DHA, one of the most important fatty acids for brain development.

3. Cut back on fish consumption to once or twice a week. Moms-to-be are often told to boost their fish consumption due to the high content of fatty acids. But it's hard to know today where your fish are coming from, and they could have high levels of mercury, PCBs, and other toxins.

4. Breast-feed if you can, as no manufactured formula can match the nutrients found in breast milk. For example, breast milk contains substances that protect a baby from diseases and infections and nurture proper growth and

development—substances that formulas don't have because they cannot be artificially synthesized. Breast-feeding has other benefits, too, such as the bonding it provides through physical contact.

A note about C-sections: C-sections do save lives, and they are medically necessary under certain situations. But only a fraction of deliveries need to be done surgically. The advantages of being born through a bacteria-filled vagina that physically baptizes the baby with life-sustaining microbes rather than a sterilized abdomen are truly stunning. Babies born via C-section face a lifetime higher risk of allergies, ADHD, autism, obesity, type 1 diabetes, and dementia later in life.

If, for whatever reason, you undergo a C-section, speak with your doctor about using the so-called gauze technique. New York University's Dr. Maria Gloria Dominguez-Bello has presented research suggesting that using gauze to collect a mother's birth canal bacteria and then rubbing the gauze over the baby's mouth and nose does help the baby grow a healthy bacterial population. It's not as good as a vaginal delivery, but it's better than a sterile C-section.

Dr. Dominguez-Bello also recommends taking probiotics and breast-feeding. She writes: "The synergy of the probiotic and prebiotic components of human breast milk provides breast-fed infants a stable and relatively uniform gut microbiome compared to formula-fed babies."

"I'VE BEEN TOLD TO TAKE ANTIBIOTICS. IS THAT OKAY?"

At some point, most of us will have to take a course of antibiotics to treat an infection. Take antibiotics only if they're absolutely necessary and recommended by your physician. Understand that antibiotics do not treat viral illnesses. Colds, the flu, and the typical sore throats that people experience are caused by viruses, and antibiotics are entirely useless.

When an antibiotic is necessary, rather than getting a "broad-spectrum" antibiotic that will kill many different bacterial species, ask your doctor for a "narrow-spectrum" medication that uniquely targets the organism that is causing the illness. And be an advocate for your children if the pediatrician wants to write a prescription for an antibiotic. Question the doctor to make sure the antibiotic is truly necessary. Antibiotics account for one-quarter of all medications for children, yet it's been shown that up to one-third of these prescriptions are not necessary.

It's important to follow your doctor's prescription exactly (i.e., do not stop taking the drug even if you feel better, as this can spur new strains of bacteria that could potentially make the situation worse). Continue to take your probiotics, but do so "on the half time," meaning take them halfway between dosages of the antibiotics. For example, if you're instructed to take the antibiotics twice daily, then take the drug once in the morning and once at night, and take your probiotics at lunchtime. And be sure to get some *L. brevis* into the mix, which is especially helpful in maintaining a healthy microbiome while taking antibiotics.

"I FEEL SO MUCH BETTER. CAN I STOP TAKING MY MEDS?"

Many people write to me to express their joy at feeling so much better after they start following my protocol. And many begin to rethink medications they are taking, wondering if they don't need them anymore. This is especially true when it comes to psychiatric drugs to treat anxiety and depression. Consider, for example, Linda T.'s experience:

> *I am fifty-two years old and am currently taking Cymbalta, 30 mg, for depression. I have been seriously depressed for too long along with severe anxiety. Only one month after becoming gluten-free and reducing my carbohydrate and sugar intake, I am a totally different person. Honestly, it is like night and day. My anxiety has disappeared; I am calm. I'm not depressed. Instead, I'm feeling good and content. I used to think that I would be on antidepressants forever, but now I feel like there's hope that one day I won't need them anymore.*

It's important that you speak with your treating physician before stopping any prescribed medication. You may indeed be able to wean yourself from certain drugs, but this should be done under the supervision of your doctor. Like Linda, have hope that you can one day say goodbye to your medication, but be smart about how you come off any meds that were prescribed to you for a reason.

A FINAL NOTE ABOUT CHILDREN

Stories of children getting their health and their future back are truly uplifting. Here is Jen W.'s story:

> *My eleven-year-old son suffered tremendously. It's no stretch to say there were days he just didn't want to wake up. He was diagnosed*

with depression, anxiety, OCD, daily nausea, severe eczema, joint pain, psychotic episodes, and unexplained weight gain. Beyond that, he was obese: 65 pounds heavier than his brother, who is thirteen months younger than him. Diets didn't work, antidepressants did nothing, and neither did the countless therapists he saw. Nothing helped, but I kept searching.

The best day of our families' lives came two years ago when a new doctor told us to eliminate sugar, gluten, dairy, processed foods, and legumes from his diet, and to go organic. By day two, all his symptoms were gone and I mean gone!!!!!! It was unreal. I am constantly telling people and doctors about my son's success! I am so so so thankful for your new book, and that there is a real movement and plan that can change people's lives.

I am routinely asked whether children can follow this protocol. You bet. In fact, children stand to gain even greater lifelong benefits, since they are still developing. I can't tell you how many parents write to me about the turnarounds they witness in their children, some of whom have been battling serious brain-related disorders, from epilepsy and ADHD to autism.

While mainstream medicine seems reluctant to embrace dietary intervention as a true medical therapy, I talk to parents all the time who report positive effects of dietary changes in their children. I would encourage any parent of a child exhibiting gastrointestinal and/or behavioral problems to try the strategies outlined in this book. The child's plate should look like yours—lots of colorful, fibrous veggies, some fruit and protein, and healthy fats.

The protocol in this book will help the vast majority of people. I'm confident that at least 80 percent of you will relieve your suffering, and all of you will be investing in your future health. Some of you, however, might need further intervention. If, after three months on this program, you do not see the results you'd hoped for, it's likely time to seek help from a practitioner trained in functional or integrative medicine. There may be deeper imbalances to address that are affecting your health and that require the support of a professional for complete healing. Don't be afraid to ask for help if you need it.

You don't need to do much to reinforce the body's instinctive proclivity toward health and optimal wellness. You are an incredibly self-regulating machine. So take a moment to appreciate—and perhaps marvel at—that wondrous reality. And then open yourself to the possibilities that await you.

PART III

LET'S EAT!

I'm a thirty-eight-year-old female and I have epilepsy. I have focal motor seizures that mimic dystonia. I suffer from night seizures and occasional daytime seizures, presenting themselves as cramping in my right arm and leg, lasting one to ten seconds. For most of my years, I believed this was a life sentence. However, I've remained mainly seizure-free on a gluten-free diet and haven't taken daily medication for years. I take muscle relaxants when I have seizures. Further, although I have remained mainly seizure-free over the last few years, I have been a terrible sleeper . . . very restless and wakeful. But I have had great success on my new diet and supplement regimen. I have never slept so good! I'm sleeping through the night and I feel so well rested in the morning.

—Anonymous from Wicklow, Ireland

CHAPTER 9

Final Reminders and Snack Ideas

CONGRATULATIONS. YOU ARE ON THE road to a better, healthier you. I'm excited for you in this next chapter of your life—one filled with a vibrancy that you might not have thought was possible before. With every meal, minute of physical movement, night of good sleep, deep exhalation of stress, and time you focus on yourself, your body is shifting in a big way and will continue to do so. All of the strategies you've learned and will eventually master will have monumental biological effects in the long term.

I predict that within the first week you'll be feeling the payoff of this protocol. You'll have fewer symptoms of a chronic condition if you currently have one, less brain fog, better sleep, and improved energy. Although you won't necessarily feel it, you'll also have a heightened protection against future illness. You'll feel stronger and more resilient. Over time, your clothes will loosen as the unwanted weight falls off, and those laboratory tests will show vast improvements in many areas of your biochemistry. For added inspiration, read Gabrielle H.'s story:

I started 2015 by adopting a gluten-free, low-carb lifestyle. I had suffered from anxiety since the age of six, and added chronic stress to that when I was in my thirties. Many other symptoms would come up later, including irritable bowel syndrome in 2005, severe joint pain/muscle aches, lack of sleep, lack of concentration, depression, and finally, a breakdown in 2009. Conventional medicine only scratched the surface. I was grasping at straws last year by trying to exclude different foods from my diet, suffering acid reflux all the while. I would fall asleep to be awakened by a dead arm and, in panic, try to get the circulation back.

This regimen has saved my life. Within five days, I started to sleep. Within twelve, I broke free from depressive moods. I have no joint pain or muscle aches now, no acid reflux at all. My gut feels like it's been repaired. I do not eat any carbs and do eat a high-fat diet with extra-virgin olive oil, coconut, butter, and the like. I feel energetic and feel that my brain is functioning again. I am sixty-three now and am looking forward to getting even better. Last week, I started eating fermented foods including kimchi, red cabbage, and cauliflower. In a week or so, I know I will find even more improvement in my overall health.

In this chapter, I'm going to set you even more firmly on your path by giving you some final reminders, a basic shopping list, and snack ideas. In the next chapter, you will find the 14-day meal plan, followed by the recipes in Chapter 11.

FINAL REMINDERS

Drink Water Throughout the Day

At a minimum, drink half of your body weight in ounces of purified water daily. If you weigh 150 pounds (68kg), drink at least 75 ounces

(2 liters) of water per day. Keep a stainless steel bottle of water with you all day. In addition, you can drink tea or coffee, and have a glass of wine with dinner. Be careful, however, about caffeine late in the day or an extra glass of wine, which will interfere with your sleep.

Avoid juicing. I realize that juice bars are all the rage today, but when you juice whole fruits and vegetables, you substantially reduce if not totally eliminate the fiber content and wind up with a sugary beverage that can rival regular soda. Don't be fooled by juice bars that advertise that juices will "cleanse" and "detoxify" your body. The same goes for sugar-filled smoothies, coconut water, and 100 percent pure watermelon water for "clean, natural hydration." Even with no added sugar, a cup of watermelon juice contains 12 grams of sugar and no fiber. If you're going to drink yourself clean, stick with filtered water.

Be Generous with Olive Oil

You are free to use olive oil liberally (extra-virgin and organic), though I trust you're not going to dump a whole cup of it onto a plate in one sitting. Note that in many cases, you can substitute coconut oil for olive oil in your recipes.

Get Used to It

For those who think "everything in moderation" makes sense, think again: "Everything in moderation" diet advice may lead to poor metabolic health. It's not the "moderation" part that's the problem; it's the "everything" part—eating every type of food available to you rather than sticking to a few choice, healthy staples.

In the 2015 Multi-Ethnic Study of Atherosclerosis, researchers looked at data from 6,814 U.S. participants—whites, blacks,

Hispanic-Americans, and Chinese-Americans. They measured diet diversity and examined how diet quality affected metabolic health. Turns out that the more diverse your diet (meaning you're eating from a very wide range of foods), the more likely you are to eat poorly and suffer metabolic consequences. In describing these results, lead researcher Dariush Mozaffarian, MD, DrPH, stated: "Americans with the healthiest diets actually eat a relatively small range of healthy foods. . . . These results suggest that in modern diets, eating 'everything in moderation' is actually worse than eating a smaller number of healthy foods."

In my experience, the healthiest people I know eat the same thing most days of the week. They have their trusty breakfasts, lunches, and dinners and don't stray from those blueprints. They generally use the same shopping lists every week. Once you've established a lineup of breakfast, lunch, and dinner ideas using my guidelines, you'll want to keep up your new patterns.

Don't Cheat

No one likes to be a cheater, but in today's world, cheating in the diet arena is practically a given. We are bombarded by choices and seductive advertising everywhere we look. Even the FDA isn't up to speed on what a "healthy" food is. In 2016 it announced plans to update its policies, recommendations, and definitions, but this will take years to execute. Believe it or not, for a long time the FDA has considered fortified cereals loaded with sugar "healthy," while it deemed avocados, salmon, and nuts "unhealthy"! It's truly ridiculous when you think about it.

Our innate survival skills might not know the difference between a slice of pizza and a slice of frittata. We have to become extremely adept at talking ourselves out of thinking one croissant won't hurt us, or that a bowl of organic mac and cheese with the kids will be okay.

This means honing a special set of survival skills, such as being hyper-aware of your brain telling you one thing ("Eat this!") while your body really needs another ("Don't eat this!"). Plan your defense when faced with temptation and obstacles to success. When a friend, for example, invites you to lunch at a restaurant where you know there's not much on the menu for you to eat, politely suggest another restaurant where you know you can live up to the principles of the Grain Brain Whole Life Plan. Don't get discouraged or let your guard down. The faster you move past these speed bumps, the healthier you'll be.

If, after following my protocol for a few weeks, you don't feel like you're getting the results you want or expect, you must ask yourself: *Am I living up to the principles of the program? Have I let things sneak into my diet through mindless nibbling or eating? Have I unwittingly caved in to the pressures of my friends offering me foods I shouldn't be eating?* This is why it's important to track your diet and record what you eat on a daily basis, especially at the start. It's also critical to make a habit of meticulously reading labels. In fact, see if you can avoid reading labels altogether by eating only foods that don't come with them!

Find a Partner

The vast majority of people who pay personal trainers do so because it buys them an accountability partner. You are somewhat forced to show up and do the work with a trainer because you've paid for that person to be there. For the same reason, it helps to have at least one other person with you on this new path. Pick someone—a friend or a family member—who wants to follow the Grain Brain Whole Life Plan with you. Work together toward your goals. Plan meals, shop, cook, and exercise as a team. Come up with new recipes and meal ideas. Share your frustrations as well as your successes along the way. After all, life is a team sport.

Make Vegetables Your Centerpiece

Stop thinking about food pyramids. Think in terms of how we eat: using a plate. A full three-quarters of your plate should be filled with fibrous, colorful, nutrient-dense whole vegetables that grow above ground. That will be your main entrée. I bet you're used to seeing your protein as the centerpiece. Now it becomes a side dish of 3 to 4 ounces (100g). Aim to consume no more than 8 ounces (225g) of protein total in a day. You'll get your fats from those found naturally in protein; from ingredients such as the butter, coconut oil, and olive oil used to prepare your meals; and from nuts and seeds (see page 198 for snack ideas).

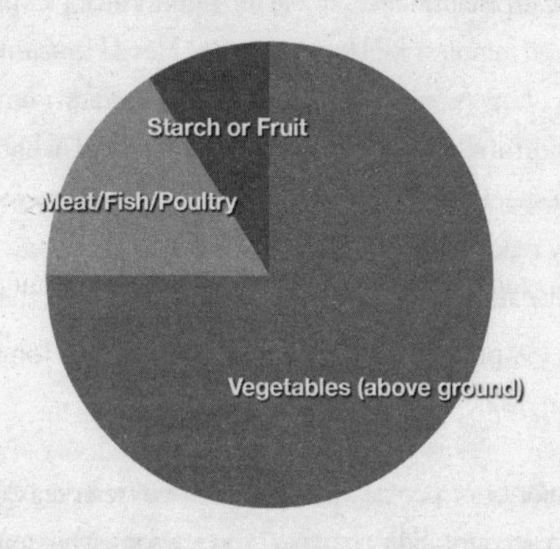

The Basic Shopping List:

- almond butter
- almond milk
- almonds
- avocado oil
- avocados
- balsamic vinegar

bell peppers

berries

black pepper

broccoli

coconut milk

coconut oil

dark chocolate

dark leafy greens, including kale and spinach

feta cheese

free-range chicken

free-range turkey

fresh guacamole

fresh salsa or pico de gallo

garlic

goat cheese

grass-fed beef

Greek-style yogurt (plain and coconut milk, 2%)

lemons

macadamia nuts

mixed greens

mozzarella

mushrooms

olive oil

olives

onions

pastured eggs

sea salt

shredded coconut

vine-ripened tomatoes

walnuts

whole fresh seasonal fruit

wild salmon

SNACKS

Better blood sugar control means you are not likely to feel super hungry in between meals. You won't be crashing an hour after that bagel breakfast because bagels are not on the program. Two 70-calorie eggs, for instance, can get you through an entire morning. So even though you might not need to snack, it's nice to know you can whenever you want to on this diet. I recommend keeping some of the less perishable snack options on hand as "emergency food" when you're on the go or at work. Keep a stash of nuts and jerky in your car, purse, or desk at work, just in case. That way, you won't get stuck running behind schedule when it's time for lunch and be tempted by the nearest fast-food restaurant or food truck. Here are some healthy snack ideas:

- a handful of raw nuts, olives, and/or seeds (no peanuts)
- a few squares of dark chocolate (anything above 70 percent cacao)
- chopped raw vegetables (e.g., bell peppers, broccoli, cucumbers, radishes) dipped into guacamole, tapenade, hummus, tahini, baba ganoush, soft goat cheese, or nut butter
- slices of cold roasted turkey, roast beef, or chicken dipped into mustard and Grain Brain Mayonnaise (page 227). [Note: Be wary of traditional deli meats, especially those that are packaged. They can be contaminated with gluten depending on how they are processed. Always ask at the deli counter for fresh, unprocessed meats that they can slice right there.]
- half an avocado drizzled with olive oil, lemon, salt, and pepper
- two hard-boiled eggs
- caprese salad: 1 sliced tomato topped with sliced fresh mozzarella cheese, drizzled with olive oil and sprinkled with basil, salt, and pepper

- Tomato-Basil Tower with Kefir Dressing, Bacon, and Fresh Dill (page 231)
- cold peeled prawns with lemon and dill
- smoked salmon (Optional: Try dipping smoked salmon in Grain Brain Mayonnaise or spreading goat cheese on top)
- one piece or serving of whole, low-sugar fruit (e.g., grapefruit, orange, apple, berries, melon, pear, cherries, grapes, kiwi, plum, peach, nectarine)
- grass-fed beef, turkey, or salmon jerky
- lacto-fermented vegetables (Try my Mixed Vegetable Kraut on page 243)
- protein bars (See my recipe online at www.DrPerlmutter.com)

Remember, this type of diet is self-regulating: You won't be victimized by blood sugar chaos brought on by too many carbohydrates stimulating irrepressible hunger and cravings. Much to the contrary, you'll feel satisfied quickly and for several hours by the fat and protein in your meals. So say goodbye to feeling foggy, sluggish, hungry, and tired throughout the day. And say hello to a whole new vibrant you. Here comes 14 days of deliciousness.

CHAPTER 10

The 14-Day Meal Plan

WELCOME TO THE MEAL PLAN, a 14-day sample menu that will serve as a model for planning your meals in the future. You'll find the recipes in Chapter 11. Although I have added nutritional data to the recipes for your information, I don't expect you to count calories and fat grams or obsess over your daily total intake. I trust you know the difference between a supersized plate of food and a reasonable portion. Vegetables that grow above ground—broccoli, asparagus, spring greens, kale, spinach, dandelion greens, cabbage, brussels sprouts, mushrooms, lettuces, leeks, radishes, bean sprouts, and cauliflower—are pretty much unlimited. You'll minimize your protein to the size of a deck of cards or the palm of your hand.

You might find it helpful to keep a food journal as you press forward. Make notes about recipes you like and foods that you don't or that could be giving you trouble (e.g., you experience symptoms such as stomach upset or headaches every time you eat sesame seeds; or you can't stand feta cheese). You can always find substitutes. Pay attention to the foods that make your body sing. You do not have to follow this menu plan strictly as written. If you love what you ate for breakfast on

Day 1, for example, feel free to repeat that again on Day 2. Don't hesitate to substitute any suggested meal for another.

Many of the dishes require that you plan (and cook) ahead. I suggest that you take a good look at these next 14 days and decide which meals you'd like to prepare. The recipes that call for fermentation time, such as the Asian-Scented Greens (page 245) and Mixed Vegetable Kraut (page 243), will need to be prepared days in advance. I purposefully didn't include them in the menu plan until the second week, so aim to make those in the first few days.

Most of these recipes serve more than one person, so think about that in your planning; you could be feeding your whole family or using leftovers the next day. In fact, on many occasions you'll simply use leftovers from the evening before to cover your lunch to make it easy and quick. And feel free to double up on recipes for a greater yield to feed more people or to have more leftovers for yourself.

Spend the day before you begin mapping out your choices, going to the market, and setting aside the time you need to spend in your kitchen. You might designate Sunday as your prep day during which you take an hour or two to get ready for the upcoming week. I recommend making a dozen hard-boiled eggs on Sunday that you can use in your meals and as snacks throughout the week.

Note that you can just as easily craft simpler meals based on the guidelines presented here (e.g., roast or steam a bounty of fresh vegetables, add a few ounces of high-quality protein, and include a mixed greens salad with lots of olive oil). For those of you who need more calories or who feel like you need more carbs, try adding additional coconut oil or olive oil first. If you still feel you need more carbs, go with non-gluten grains such as quinoa or wild rice and keep the portion small (8 tablespoons). Or try my Sunchoke Fritters (page 234), which are an excellent substitute for the carbs (e.g., potatoes, bread, pasta) you will not be eating.

When preparing a salad of mixed greens, be sure to add lots of cooked or raw vegetables (see below). You might not want to eat raw asparagus and broccoli, for instance, but you can cook those veggies and add them to your leafy green salad with cucumbers, jicama, radishes, and such. I don't suggest which type of dressing to use in all instances. A fallback staple is olive oil and balsamic vinegar. Avoid processed, commercially made dressings. These can be loaded with sugar, bad fats, fillers, and artificial ingredients. Read your labels.

You don't need to toss your cookware, cutting boards, utensils, and so on that have been used with gluten-containing products in the past. Use what you've already got. Think about investing in some fun kitchen equipment in the future to make the art of cooking all the more enjoyable.

Get to know your grocers and local farmer's markets; the people there can tell you what just came in and where your foods are coming from. Go organic, grass-fed, and wild whenever possible. Aim to choose produce that's in season in your hemisphere, and be willing to try new foods you've never had before. Take one to three breaths with your eyes closed before you eat and express gratitude for the food that will nourish you from the inside out.

The most essential lesson to learn as you commence this new way of eating (and living!) is to begin to listen to your body. It knows what it needs. When we clear the slate of processed, inflammatory foods, we begin to guide ourselves toward our best selves.

14 DAYS OF DELICIOUSNESS

Recipes in boldface are included in Chapter 11. Recipes marked with an asterisk (*) can be found on my website, www.DrPerlmutter.com.

Day 1:

Breakfast: 2 poached eggs topped with salsa or pico de gallo + half an avocado drizzled with olive oil and a pinch of sea salt

Lunch: **Layered Vegetable Salad** (page 226) with 75g diced grilled chicken mixed in

Dinner: **Mixed Greens with Toasted Walnuts** (page 228) + 75g baked or grilled fish

Dessert: 2 squares of dark chocolate dipped in 1 tablespoon almond butter

Day 2:

Breakfast: skip!

Lunch: **Onion Soup** (page 220) + 2 roasted chicken drumsticks + side of mixed greens

Dinner: **Tuscan-Style Pork Roast** (page 255) + sautéed spring greens + 100g quinoa (optional)

Dessert: **Coconut Pudding** (page 258)

Day 3:

Breakfast: **Broccoli, Mushroom, and Leek Frittata** (page 210) + 250ml almond milk

Lunch: leftover pork roast tossed into mixed greens salad with at least 3 raw vegetables (e.g., broccoli, radishes, green beans) + half an avocado + drizzle of olive oil

Dinner: grilled steak + roasted vegetables + leftover onion soup

Dessert: 60g fresh berries topped with coconut milk

Day 4:

Breakfast: leftover frittata + 250ml almond milk (optional)

Lunch: mixed greens salad with at least 3 raw or cooked vegetables topped with grilled fish or chicken

Dinner: **Roast Leg of Grass-Fed Lamb** (page 253) + unlimited steamed vegetables + 100g wild rice (optional)

Dessert: **Ricotta with Berries and Toasted Almonds** (page 260)

Day 5:

Breakfast: **Strawberry Power Smoothie** (page 213)

Lunch: leftover lamb tossed into mixed greens salad with at least 3 raw vegetables (e.g., broccoli, radishes, green beans) + half an avocado + drizzle of olive oil

Dinner: **Steamed Wild Salmon with Sautéed Leeks and Chard** (page 248) + 100g rice or quinoa (optional)

Dessert: skip!

Day 6:

Breakfast: Greek-style yogurt topped with raw walnuts and fresh berries

Lunch: mixed greens salad with 2 hard-boiled eggs, at least 3 raw vegetables (e.g., celery, spring onions, water chestnuts), half an avocado, crushed walnuts, and shredded or diced cheddar cheese + 1 piece whole fruit

Dinner: grilled fish, chicken, or steak + grilled courgettes + **Braised Kale** (page 233)

Dessert: **Easy Chocolate Mousse** (page 259)

Day 7:

Breakfast: **Baked Eggs and Greens** (page 212)

Lunch: mixed greens and vegetables sautéed in butter and garlic + grilled chicken or fish

Dinner: **Lamb Meatball Soup** (page 223)

Dessert: 2 or 3 squares dark chocolate

Day 8:

Breakfast: 2 fried eggs topped with diced avocado, diced vine-ripened tomatoes, and drizzle of olive oil + unlimited sautéed greens and other vegetables

Lunch: Leftover lamb meatball soup

Dinner: **Broccoli, Mushroom, and Feta Toss** (page 242) + **Herb-Roasted Wild Salmon** (page 247)

Dessert: whole fruit

Day 9:

Breakfast: 3 scrambled eggs with at least 3 vegetables (e.g., spinach, mushrooms, onions) and goat cheese + 250ml almond milk (optional)

Lunch: **Jicama Salad** (page 230) + side of roasted turkey

Dinner: **Thai Vegetable Curry** (page 239) + 75–115g ounces chicken or steak
Dessert: **Coconut Pudding** (page 258)

Day 10:

Breakfast: coconut milk or plain yogurt topped with nuts and seeds + 2 hard- or soft-boiled eggs
Lunch: **Layered Vegetable Salad** (page 226) with 75g diced grilled chicken mixed in
Dinner: **Roasted Chicken Thighs with Parsley Sauce** (page 256) + unlimited steamed vegetables + 100g quinoa (optional)
Dessert: skip!

Day 11:

Breakfast: Eggs Benedict with Courgette Pancakes* or **Breakfast "Porridge"** (page 214)
Lunch: leftover roasted chicken thighs + mixed greens and vegetable salad
Dinner: grilled fish of your choice + roasted asparagus and brussels sprouts + **Sunchoke Fritters** (page 234)
Dessert: 25–50g cheese

Day 12:

Breakfast: skip!
Lunch: mixed greens salad with at least 3 raw or cooked

vegetables topped with grilled fish or chicken + **Sunchoke Gratin** (page 236)

Dinner: grilled chicken or fish + **Mixed Vegetable Kraut** (page 243)

Dessert: **Easy Chocolate Mousse** (page 259)

Day 13:

Breakfast: coconut milk yogurt topped with nuts and seeds + 2 hard- or soft-boiled eggs

Lunch: leftover vegetable kraut tossed into mixed greens salad or beside grilled fish or poultry

Dinner: **Grass-Fed Beef Burgers** (page 252) + mixed green salad or **Fish Fillets with Black Olives, Artichokes, and Shaved Brussels Sprout Slaw** (page 250)

Dessert: skip!

Day 14:

Breakfast: **Strawberry Power Smoothie** (page 213) or Oatless Oatmeal* + 2 eggs any style

Lunch: **Creamy Cauliflower Soup** (page 221) + mixed greens salad with shredded chicken mixed in

Dinner: **Pea and Goat Cheese Custards** (page 218) + mixed greens salad + 75g meat or fish

Dessert: whole fruit

Congratulations! You've made it through two weeks on the Grain Brain Whole Life Plan eating nutrient-dense foods that fill your heart and soul. Hopefully you've incorporated other elements on that checklist (see page 173) into your new lifestyle, too. I'm confident you can keep going. If you don't know what to eat after these two weeks, just repeat the same 14-day meal plan until you get used to cooking and eating this way and feel confident enough to start experimenting in the kitchen. Now let's get to the recipes.

CHAPTER 11

The Recipes

GET READY TO MAKE SOME delicious meals using the recipes in this chapter. When buying ingredients, remember to choose organic, grass-fed, GMO-free, gluten-free, and wild whenever possible. Reach for extra-virgin olive and coconut oils. Check labels on all packaged goods to be sure they don't contain anything suspicious (see page 88). Most of the ingredients you'll need are now widely available and found in supermarkets. Some of these recipes are more time-consuming to make than others, so plan ahead and feel free to swap one for another if you don't have the extra time. Ultimately, have fun with these recipes and enjoy being your own personal chef.

EGGS AND OTHER BREAKFAST DISHES

Broccoli, Mushroom, and Leek Frittata

Serves 4

A frittata can be made with almost any combination of vegetables and/or meat, even leftovers. Some tasty combinations might be pumpkin-mint, tomato-basil, asparagus-salmon, onion–chopped greens, summer squash–feta, chopped pork–Gruyère—the list can go on and on. Frittatas are perfect for breakfast, brunch, lunch, or dinner and can be eaten hot out of the oven or at room temperature.

1 tablespoon unsalted butter, preferably from grass-fed cows

1 tablespoon extra-virgin olive oil

225g diced leeks, white part only

6 large mushrooms, stems removed, cleaned, and thinly sliced

1 teaspoon minced garlic

120g finely chopped tenderstem broccoli or broccoli

Sea salt and freshly ground black pepper

5 large eggs

4 tablespoons grated Parmesan cheese

2 large egg whites

Preheat the oven to 180°C/gas 4.

Generously butter a 20cm deep-dish pie plate or ovenproof frying pan. Set aside.

Combine the butter and olive oil in a large sauté pan over medium heat. Add the leeks and cook, stirring frequently, for about 4 minutes or just until wilted. Add the mushrooms and garlic and continue to cook, stirring frequently, for about 12 minutes or until the mushrooms have exuded their liquid and begun to brown. Stir in the broccoli and continue to cook, stirring frequently, for another 3 to 4 minutes, until the broccoli is slightly soft. Lightly season with salt and pepper.

While the vegetables are cooking, place the whole eggs in a medium bowl, whisking to lighten. Add 2 tablespoons of the cheese and season with salt and pepper.

Place the egg whites in a medium bowl and, using a handheld electric mixer, beat until firm, but not dry. Fold the beaten egg whites into the egg mixture, folding only until small pieces of egg white are still visible.

Scrape the broccoli mixture into the eggs, stirring to blend. Pour into the prepared pan, gently smoothing out the top with a spatula. Sprinkle with the remaining 2 tablespoons of cheese and transfer to the oven.

Bake for about 20 minutes or until the center is set and the top is golden brown and almost crisp around the edges.

Remove from the oven and let stand for a couple of minutes before cutting into wedges and serving.

Nutritional Analysis per Serving: calories 278, fat 15 g, protein 18 g, carbohydrates 20 g, sugar 6 g, fiber 6 g, sodium 286 mg

Baked Eggs and Greens

Serves 6

This is a terrific Sunday brunch dish. The recipe is easily doubled; just use two baking dishes. Make sure that you remove the dish from the oven before the eggs are fully cooked, as you want the yolks to be runny when served so that they can be mixed in with the greens.

- 1 tablespoon extra-virgin olive oil
- 1 tablespoon unsalted butter, preferably from grass-fed cows
- 115g chopped leeks, white part only
- 1 tablespoon chopped garlic
- Sea salt and freshly ground black pepper
- 2 bunches Swiss chard, tough stem ends removed and cut into large pieces
- 4 tablespoons chopped sun-dried tomatoes
- 1 tablespoon chopped fresh basil
- 75ml double cream, preferably from grass-fed cows
- 12 large eggs
- 60g grated fontina cheese

Preheat the oven to 200°C/gas 6.

Generously coat a 23 by 32 by 5cm baking dish with butter. Set aside.

Heat the oil and butter in a large frying pan over medium heat. Add the leeks and garlic, season with salt and pepper, and cook, stirring occasionally, for about 8 minutes or until the leeks are quite soft.

Begin adding the chard, a couple of handfuls at a time, tossing to soften and wilt before adding another batch. When all of the chard has been added, add the tomatoes and basil. Season with salt and pepper and continue to cook, tossing and turning, for about 10 minutes or until very soft.

Stir in the cream and continue to cook for about 6 minutes or until the cream has almost evaporated. Taste and, if necessary, season with additional salt and pepper.

Spoon the chard mixture into the prepared baking dish, spreading it out into an even layer. Using the back of a soup spoon, make 12 small indentations in the chard. Crack one egg into each indentation. When all of the eggs are nestled in the chard, season each one with salt and pepper and sprinkle the cheese over the top, covering both eggs and chard.

Transfer to the oven and bake for about 15 minutes or until the whites are not quite firm and the yolks are still very runny.

Remove from the oven and let stand for 5 minutes to allow the whites to set before serving.

Nutritional Analysis per Serving: calories 297, fat 21 g, protein 17 g, carbohydrates 10 g, sugar 3 g, fiber 3 g, sodium 585 mg

Strawberry Power Smoothie

Serves 1

Most traditional smoothies and shakes are filled with sugar, but this one lives up to my standards and is an excellent recipe to have on hand for those mornings when you don't have time to create a regular breakfast meal. This smoothie can go with you to work, too, and keep you satisfied for hours.

- 4 tablespoons unsweetened coconut milk
- 4 tablespoons water (or more for desired consistency)
- 40g frozen strawberries
- ¼ ripe avocado, pitted and peeled
- 1 tablespoon raw unsalted sunflower seeds or almonds
- 1 tablespoon hemp seeds
- 1 tablespoon sunflower seed butter or almond butter
- 1 teaspoon chopped fresh ginger
- ½ teaspoon ground cinnamon

Combine all of the ingredients in a blender jar. Blend until completely smooth, scraping down the sides as needed. Serve immediately.

Nutritional Analysis per Serving: calories 380, fat 32 g, protein 10 g, carbohydrates 17 g, sugar 7 g, fiber 7 g, sodium 23 mg

Breakfast "Porridge"

Serves 1

Once you taste this bowl of deliciousness, you'll never want to go back to your old-fashioned oatmeal. To accompany this dish, have a cup of coffee or kombucha tea, or drink some kefir, almond milk, or coconut milk. This breakfast will keep you satisfied all morning long.

125ml hot water (or more for desired consistency)

1½ tablespoons chia seeds

1½ tablespoons hemp seeds

1–2 tablespoons sunflower lecithin (optional)

1 tablespoon coconut oil

1 tablespoon almond butter

1 teaspoon ground flax seeds (optional)

1 teaspoon ground cinnamon

5 drops stevia, or to taste

Sea salt

60g blueberries, raspberries, and/or blackberries

Combine all of the ingredients except for the berries in a bowl. Stir well. Top with the berries and serve.

Nutritional Analysis per Serving: calories 460, fat 37 g, protein 12 g, carbohydrates 26 g, sugar 9 g, fiber 11 g, sodium 330 mg

APPETIZERS

Wild Salmon Crudo with Shaved Artichokes

Serves 4

This light but beautiful artichoke salad is the perfect complement to the unctuous raw salmon. If you can't find tender baby artichokes, the salad can be made with raw asparagus or thinly sliced fennel. Eating both the salmon and the artichokes raw gives you many health benefits.

- 225g wild salmon fillet, skin and pin bones removed
- 75ml white vinegar
- 3 baby artichokes
- 4 tablespoons extra-virgin olive oil, plus more as necessary
- 2 teaspoons fresh lemon juice
- 2 tablespoons chopped fresh chives, tarragon, or flat-leaf parsley
- Sea salt and freshly ground black pepper
- Lemon wedges, for garnish (optional)

Using a very sharp knife, cut the salmon crosswise against the grain into 5mm-thick slices. Place an equal number of the slices in a single layer on each of four chilled plates.

Cover each plate with a sheet of plastic wrap. Working with one plate at a time and using the bottom of a small frying pan (or any flat object), gently press down to flatten the salmon so that it covers the entire plate. Do not press too hard, as you don't want to make the salmon mushy. Leaving the plastic wrap on, transfer the plates to the refrigerator.

Fill a large bowl with cold water. Add the white vinegar and set aside.

Working with one at a time, pull off the tough outer leaves of each artichoke. Then, using kitchen scissors, cut off the pointed, spiked tips of each artichoke and about 5mm off of the top. If the artichokes have stems, cut them off completely also.

Using a vegetable slicer or very sharp knife, slice each artichoke crosswise into paper-thin slices. Immediately drop the slices into the cold, acidulated water to keep the flesh from oxidizing.

When all of the artichokes have been sliced, remove them from the water and pat very dry. Place the well-drained slices in a medium bowl and add 2 tablespoons of the olive oil, along with the lemon juice. Add the herbs, season with salt and pepper, and toss to coat.

Remove the fish from the refrigerator and unwrap. Drizzle an equal portion of the remaining 2 tablespoons olive oil over each plate. Season lightly with salt and pepper. Scatter an equal portion of the shaved artichokes over each plate. If using, garnish with a lemon wedge, and serve immediately.

Nutritional Analysis per Serving: calories 260, fat 17 g, protein 17 g, carbohydrates 13 g, sugar 2 g, fiber 6 g, sodium 260 mg

Pea and Goat Cheese Custards

Serves 4

This very elegant dish can be served as a first course for a dinner party or as a lovely lunch when accompanied by a green salad. Although it's quite rich, the crisp peas and fresh herbs add an unexpected lightness.

Butter, for ramekins

150g frozen petits pois

75g mild creamy goat cheese

4 extra-large eggs, at room temperature

250ml double cream, preferably from grass-fed cows

2 tablespoons grated Parmesan cheese

Sea salt and freshly ground black pepper

2 tablespoons finely chopped spring onions white part only

2 tablespoons minced fresh dill

4 sprigs fresh dill, for garnish (optional)

Preheat the oven to 180°C/gas 4.

Generously butter the interior of four 175ml ramekins. Set aside.

Bring a small pot of water to a boil, and boil the peas for 1 minute. Drain well and pat dry. Set aside.

Place the goat cheese in the bowl of a food processor fitted with the metal blade. Add the eggs, cream, and Parmesan. Season with salt and pepper and process to a smooth puree.

Scrape the cheese mixture into a medium bowl. Add the spring onions and minced dill and stir to blend well.

Season the peas with salt and pepper and spoon an equal portion into the bottom of each of the buttered ramekins. Then spoon an equal portion of the cheese mixture over the peas.

Place the ramekins in a baking pan. Add enough hot water to come halfway up the sides of the ramekins and carefully transfer the baking pan to the oven.

Bake for 25 minutes or until the custards are set in the center and lightly browned around the edges.

Remove from the oven and place the ramekins on a wire rack to cool for 10 minutes.

Garnish each ramekin with a dill sprig, if desired, and serve while still very warm.

Nutritional Analysis per Serving: calories 390, fat 34 g, protein 14 g, carbohydrates 8 g, sugar 1 g, fiber 2 g, sodium 370 mg

SOUPS

Onion Soup

Serves 6

This is about as rich and delicious as the classic French onion soup, even though it lacks the traditional topping of toasted baguette. Although it can be made with all red or all sweet onions, the combination creates a rich color and a slightly sweet flavor.

115g unsalted butter, preferably from grass-fed cows
550g thinly sliced red onions
550g thinly sliced sweet onions
2 bay leaves
1 star anise
125ml brandy
2 liters beef stock or low-sodium beef broth
Sea salt and freshly ground black pepper
175g grated Gruyère cheese

Place the butter in a large saucepan over medium-low heat. Add the onions, bay leaves, and star anise and cook, stirring frequently, for about 20 minutes or until the onions have begun to caramelize and turn a rich golden brown.

Stir in the brandy, raise the heat, and bring to a boil. Boil for 3 to 4 minutes so that the alcohol can cook off. Add the stock and season with salt and pepper. Bring to a boil, then lower the heat and cook at a gentle simmer for 30 minutes or until the onions are meltingly soft and the soup is deeply flavored. Remove and discard the bay leaves and star anise.

Taste and, if necessary, adjust the seasoning. Ladle into deep soup bowls and immediately top each bowl with an equal portion of the cheese so that the heat can begin melting it.

Serve immediately.

Nutritional Analysis per Serving: calories 360, fat 24 g, protein 14 g, carbohydrates 15 g, sugar 9 g, fiber 2 g, sodium 370 mg

Creamy Cauliflower Soup

Serves 4

Although called "creamy," this soup does not have an ounce of cream in it. It is pure velvety vegetable flavor that is perfection in a bowl. The addition of the brown butter adds extraordinary richness to what is otherwise a very simple soup.

The soup may be made up to 2 days in advance and stored in the refrigerator in an airtight container. Reheat it and make the brown butter just before serving.

1 head cauliflower, trimmed and broken into small pieces, including the tender core

115g chopped leeks, white part only

Sea salt

115g unsalted butter, preferably from grass-fed cows

Freshly ground white pepper

Reserve 8 tablespoons of the cauliflower pieces and place the remaining cauliflower in a medium saucepan. Add the leeks, along with 1 liter cold water. Generously season with salt and place over medium-high heat. Bring to a boil, then cover and simmer for about 12 minutes or until the cauliflower is very soft.

While the cauliflower is cooking, heat the butter in a small frying pan over medium-low heat. Add the reserved cauliflower pieces and sauté, stirring frequently, for about 7 minutes or until the butter is golden brown with a nutty aroma and the cauliflower is lightly browned and just barely cooked. Remove from the heat and keep warm.

Remove the cauliflower and leek mixture from the heat and, using a slotted spoon, transfer the vegetables to a blender jar or the bowl of a food processor fitted with the metal blade. Add 250ml of the cooking water and reserve the remaining water.

With the motor running, begin pureeing the cauliflower, slowly adding additional cooking water until the mixture reaches a soup-like consistency. Season with salt and white pepper.

Ladle an equal portion of the soup into each of four large shallow soup bowls. Spoon a dollop of the sautéed cauliflower in the center of each bowl and drizzle an equal portion of the brown butter over the top.

Serve immediately.

Nutritional Analysis per Serving: calories 240, fat 23 g, protein 3 g, carbohydrates 8 g, sugar 3 g, fiber 3 g, sodium 314 mg

Lamb Meatball Soup

Serves 8

This dish comes courtesy of Seamus Mullen, chef-proprietor of Tertulia restaurant in New York City. It's a great recipe to use for dinner parties or to make on a Sunday night. Use the leftovers for lunches in the upcoming week.

For the meatballs:

2 large eggs

135g almonds, soaked in milk for 30 minutes, then drained and finely chopped

8 tablespoons chopped mixed fresh herbs, such as oregano, rosemary, and/or thyme

1 tablespoon red wine (optional)

2 cloves garlic, minced

2 tablespoons sea salt

1 teaspoon cayenne pepper

1 teaspoon ground coriander

1 teaspoon ground cumin

1 teaspoon ground fennel

½ teaspoon freshly ground black pepper

1.1kg minced lamb

For the soup:

2 tablespoons extra-virgin olive oil, plus more for garnish

1 bunch (4 to 6) small carrots, chopped

4 cipollini onions or shallots, peeled

100g diced king oyster mushrooms

1 fennel bulb, trimmed and cut into 2.5cm pieces

2 cloves garlic, sliced

250ml white wine

1.5 liters chicken stock

2 bay leaves

2 sprigs fresh thyme

1 sprig fresh rosemary

Sea salt and freshly ground black pepper

170g red quinoa, rinsed

1 jalapeño pepper, stemmed, seeded, and sliced as thinly as possible

200g sugar snap peas, cut diagonally in half

40g coarsely chopped radicchio

Freshly chopped dill, coriander, basil, fennel fronds, and/or mint, for garnish

For the meatballs: Whisk the eggs in a large bowl. Add all of the remaining meatball ingredients except for the lamb and mix thoroughly. Add the lamb, then, using your hands, blend everything together. Pinch off a piece of the lamb mixture and gently roll between your hands to form 4cm balls. Continue shaping until all the meat mixture is used.

For the soup: In a large saucepan, heat the olive oil over high heat and quickly brown the meatballs evenly. Transfer to a plate lined with paper towels. Add the carrots, onions, mushrooms, and fennel to the saucepan, and sauté for 3 minutes, then add the garlic and cook for

1 minute. Deglaze with the white wine and allow the alcohol to cook off, about 3 minutes. Add the chicken stock, bay leaves, thyme, and rosemary, and bring to a boil. Reduce the heat to a simmer, and season with salt and pepper.

Add the quinoa and simmer for 15 minutes, until it's just tender, then add the meatballs and gently simmer for 2 minutes. Check the meatballs for an internal temperature of about 48°C; if touched to your lower lip, they should be warm, but not superhot. Once they have reached 48°C at the center, add the jalapeño, sugar snap peas, and radicchio. Simmer for another 3 minutes or until the vegetables are just barely tender but still vibrant.

Serve immediately, finishing each bowl with a healthy drizzle of olive oil and a generous sprinkling of chopped herbs.

Nutritional Analysis per Serving: calories 650, fat 35 g, protein 40 g, carbohydrates 45 g, sugar 8 g, fiber 13 g, sodium 680 mg

SALADS

Layered Vegetable Salad

Serves 6

This is a terrific salad to make when company is expected, as it can be made ahead and tossed at the last minute. The red onions add some nice color, but if you don't have them on hand, white onions will work just fine. However, don't replace the softer savoy or chinese cabbage with ordinary green or red cabbage, as the latter is a bit too tough.

3 red onions, peeled and trimmed

675g thinly sliced savoy cabbage or chinese leaves

1 large jicama or celery heart, peeled, trimmed, and shredded

500g thinly sliced radish, preferably red, but any type will do

125ml organic, cultured, full-fat plain yogurt

125ml Grain Brain Mayonnaise (recipe follows)

2 tablespoons chopped sustainably sourced anchovies packed in olive oil (see next recipe for details)

2 teaspoons chopped mixed fresh herbs, such as mint, basil, parsley, and/or thyme

Sea salt and freshly ground black pepper (optional)

Using a vegetable slicer or a very sharp knife, cut the onions crosswise into paper-thin slices. Place the slices in a large bowl of ice water and let soak for 10 minutes. Pour off the water and pat the onions dry.

Place a thin layer of cabbage in the bottom of a large salad bowl. Top with a thin layer of onions, followed by the jicama, and finally the radishes. Continue making thin layers, ending with radishes, until all of the vegetables have been used.

Combine the yogurt, mayonnaise, anchovies, and herbs in a small bowl and whisk until well combined. Pour the dressing over the salad, spreading it evenly over the top. Cover with plastic wrap and refrigerate for at least 6 hours or up to 24 hours.

When ready to serve, toss the salad. Taste and, if desired, season with salt and pepper.

Nutritional Analysis per Serving: calories 232, fat 16 g, protein 5 g, carbohydrates 17 g, sugar 7 g, fiber 6 g, sodium 390 mg

Grain Brain Mayonnaise

Makes about 500ml

The secret to this mayonnaise is in its oil. Rather than traditional mayo, which typically uses rapeseed oil, this one calls for avocado oil, which creates a much more delicious, nutritious experience. Use this mayo like you would traditional mayo—as a spread, as a dip, and in dressings. Be sure to buy organic avocado oil. Sustainably sourced anchovies are available online.

3 large egg yolks, at room temperature

½ teaspoon sea salt

¼ teaspoon mustard powder

1 tablespoon champagne vinegar or fresh lemon juice

375 – 500ml avocado oil

1 tablespoon hot water

Fill a blender jar with boiling water and set it aside for a couple of minutes. You just need to heat the jar to help the eggs thicken. Pour out the water and quickly wipe the jar dry. Place the jar on the blender motor. Add the egg yolks and process on medium speed until very thick. Add the salt and mustard powder and quickly incorporate. Add the vinegar and process to blend.

With the motor running, begin adding the oil in an excruciatingly slow drip; the slower the drip, the more even the emulsification. When about half of the oil has been added, you should have a sauce that is like double cream, and you can then begin adding the oil just a bit quicker, as curdling will no longer be an issue. If the mixture seems to be too thick—you want a soft, creamy mix—add just a smidge more vinegar. Continue adding the oil until all of it has been absorbed into the eggs. Then, add just enough hot water (but no more than 1 tablespoon) to smooth the mix. Scrape the mayonnaise into a clean container with a lid. Cover and refrigerate for up to 5 days.

Nutritional Analysis per Serving (1 tablespoon): calories 105, fat 11 g, protein 0 g, carbohydrates 0 g, sugar 0 g, fiber 0 g, sodium 34 mg

Mixed Greens with Toasted Walnuts

Serves 4

Caramelized onion dressing, crunchy walnuts, and slightly bitter greens combine to make an aromatic and satisfying salad. It works well as a lunch main course or as a side dish for grilled fish or poultry.

- 1 large red onion, peeled and cut lengthwise into 8 wedges
- 125ml cup plus 1 tablespoon walnut oil
- 1 tablespoon balsamic vinegar
- 4 tablespoons chicken stock or canned low-sodium chicken broth
- 3 tablespoons white balsamic vinegar
- Sea salt and freshly ground black pepper
- 500g finely chopped mixed bitter greens, such as endive, radicchio, dandelion, mustard, and/or kale
- 115g chopped toasted walnuts
- 1 small red shallot, peeled, cut in half lengthwise, and thinly sliced

Preheat the oven to 200°C/gas 6.

Place the onion wedges, cut side down, in a nonstick baking pan. Combine 1 tablespoon of the oil with the balsamic vinegar and drizzle the mixture over the onions. Transfer to the oven and roast, turning occasionally, for about 30 minutes or until golden brown and caramelized.

Remove from the oven and set aside to cool slightly. You want the onions to still be quite warm when you make the dressing.

While still warm, place the seasoned onions in a food processor fitted with the metal blade. Add the remaining 125ml oil along with the stock and the white balsamic vinegar. Process to a smooth, thick puree. Season with salt and pepper. (The dressing may be made in advance, but if so, you will need to warm it slightly before adding to the salad.)

Place the greens in a large salad bowl. Pour the dressing over the top, adding just enough to coat and wilt the greens. You may not need all of the dressing. Toss well.

Add the toasted walnuts and shallot and again toss to combine. Taste and, if necessary, season with additional salt and pepper.

Serve immediately.

Nutritional Analysis per Serving (if all the dressing is used): calories 600, fat 53 g, protein 14 g, carbohydrates 30 g, sugar 5 g, fiber 17 g, sodium 140 mg

Jicama Salad

Serves 4

The zesty dressing is a perfect match for the slightly sweet, crisp jicama. When paired with the bitter radicchio, it makes for a salad of complex textures and savory flavors.

4 tablespoons finely chopped sun-dried tomatoes
1 tablespoon chopped fresh coriander
1 tablespoon chopped fresh chives
3 tablespoons champagne vinegar
2 teaspoons fresh lime juice
1 teaspoon fresh lemon juice
2 teaspoons extra-virgin olive oil
Freshly ground black pepper
350g julienned jicama
40g shredded radicchio
Parmesan or ricotta salata cheese, for shaving

Combine the tomatoes with the coriander and chives in a small non-reactive container. Stir in the vinegar, along with the lime and lemon juices and the olive oil. Season generously with pepper and stir to blend well. Cover and refrigerate for at least 1 hour or up to 4 hours.

Place the jicama in a large bowl of ice water and refrigerate for 1 hour.

When ready to serve, drain the jicama very well and pat dry. Place in a medium bowl and pour the tomato dressing over the top. Toss to blend well.

Place a layer of radicchio in the center of each of four salad plates. Mound an equal portion of the jicama salad in the center of each plate. Shave the cheese over each plate, and serve immediately.

Nutritional Analysis per Serving: calories 180, fat 9 g, protein 10 g, carbohydrates 12 g, sugar 3 g, fiber 5 g, sodium 350 mg

Tomato-Basil Tower with Kefir Dressing, Bacon, and Fresh Dill

Serves 1

This recipe is from my good friend Fabrizio Aielli, chef at Sea Salt restaurant in my hometown of Naples, Florida. Enjoy it as an appetizer, a refreshing snack on the weekend, or a side to a meal. When you find that perfectly ripe tomato, make this recipe.

1 vine-ripened tomato, sliced into 3 slices, top and bottom discarded
2 fresh basil leaves
2 tablespoons Kefir Dressing (recipe follows)
2 slices bacon, cooked until crispy and finely chopped
1 tablespoon extra-virgin olive oil
Sea salt

Stack the tomato slices on a plate, tucking a basil leaf between each layer. Drizzle with the dressing, sprinkle the bacon on top, and finish with the olive oil and salt.

Nutritional Analysis per Serving: calories 273, fat 24 g, protein 9 g, carbohydrates 9 g, sugar, 6 g, fiber 2 g, sodium 480 mg

Kefir Dressing

Makes about 500ml

Kefir has a tart and refreshing flavor. Its texture is similar to a drinking-style yogurt, so it makes a great dressing.

- 500ml kefir
- 2 tablespoons red wine vinegar
- 1 sprig fresh dill, chopped
- 2 tablespoons extra-virgin olive oil
- Sea salt and freshly ground black pepper

In a medium bowl, whisk together the kefir, vinegar, and dill. Whisk in the olive oil, a little at a time, until it is fully incorporated. Season with salt and pepper. Store in an airtight container in the refrigerator for up to 1 week.

Nutritional Analysis per Serving (2 tablespoons): calories 34, fat 3 g, protein 1 g, carbohydrates, 1 g, sugar 1 g, fiber 0 g, sodium 50 mg

VEGETABLES

Braised Kale

Serves 4

After years of being ignored, except in Portuguese cuisine, kale is having its day in the sun. It is high in fiber, filled with antioxidants and vitamins, and a great detoxifier. It has been shown to help lower the risk of many cancers. I think that this is a particularly delicious recipe to add to your kale repertoire.

2 bunches lacinato (or other type) kale
3 tablespoons extra-virgin olive oil
1 large sweet onion, peeled, trimmed, and cut into slivers
1 tablespoon (about 5 large cloves) roasted garlic puree (see Note)
Sea salt
Red chilli flakes
2 tablespoons red wine vinegar

Trim off the tough lower stems of the kale. Stack the leaves and cut them crosswise into thick pieces. Wash very well in cold water, taking care that all dirt has been rinsed off. Drain well, but do not spin off all of the water, as you need it to make the braising liquid.

Heat the oil in a large deep sauté pan over medium heat. Add a layer of kale, along with the onion slivers, and let wilt; then continue adding kale and tossing to incorporate until all of the kale has been added to the pan. Toss in the garlic puree and season with salt and chilli flakes. Cover and braise for about 10 minutes or until very, very tender.

Remove from the heat and uncover. Drizzle in the vinegar and toss to blend. Serve immediately.

NOTE: To make roasted garlic puree: Preheat the oven to 180°C/gas 4. If roasting whole heads, lay the head on its side and, using a sharp knife, cut about 3mm off the stem end. Lightly coat the entire head(s) or cloves of garlic with extra-virgin olive oil. Wrap tightly in parchment paper and place in a baking pan in the oven. Roast until soft and aromatic; whole heads should take about 25 minutes and individual cloves about 12 minutes. Remove from the oven, unwrap, and let cool slightly. Using your fingertips, squeeze the flesh from the skin. The cloves may or may not pop out whole, but either way, it doesn't matter, as roasted garlic usually gets mashed or pureed before use. Use immediately, or cover and refrigerate for up to 1 week.

Nutritional Analysis per Serving: calories 210, fat 12 g, protein 6 g, carbohydrates 24 g, sugar 9 g, fiber 6 g, sodium 140 mg

Sunchoke Fritters

Serves 4

Sunchokes, also known as Jerusalem artichokes, have nothing to do with artichokes or Jerusalem, although they do have a flavor profile quite similar to artichokes. Although often eaten raw in salads, when cooked they can be used in place of potatoes, as they are in this recipe, which is similar to the traditional Jewish dish latkes.

- 900g Jerusalem artichokes, scrubbed and dried
- 1 shallot, peeled and minced
- 4 tablespoons unsalted butter, preferably from grass-fed cows, melted, plus more as needed
- Sea salt and freshly ground black pepper

Using a vegetable shredder, cut the artichokes into julienne. (Alternatively, you may use a food processor fitted with the shredding blade, but this will create wetter shreds than the small, drier strips you desire.) Place the strips in a medium bowl and toss them with the shallot.

Place the mixture into a clean kitchen towel. Pull up the sides of the towel and tightly twist it closed. Continue tightening as you press out any excess liquid from the vegetables.

Add 2 tablespoons of the butter to a 25cm nonstick frying pan placed over low heat. Add the drained artichoke mixture, patting it down with a spatula to make a dense cake. Season with salt and pepper. Cook over low heat for about 12 minutes or until the bottom is crisp and golden brown. Adjust the heat as necessary so that the cake does not get too dark before the artichoke begins to cook through. Add more butter as necessary to keep the cake from sticking.

If you feel like living dangerously, lift the cake up from the pan and turn it over using two spatulas. If not, slide the cake out onto a plate. Then use a second plate inverted over the cake and carefully turn the plate upside down so that the whole cake flips around, browned side up, and then slide it back into the pan.

Return the pan to low heat and again pat the mixture down into the pan. Drizzle the remaining 2 tablespoons butter around the edge of the pan and continue to cook for another 7 minutes or until golden brown and crisp on the bottom and the artichoke are cooked through.

Place a double layer of paper towels on a clean, flat surface. Gently tip the cake onto the paper towels and let rest for a minute or so to allow some of the excess butter to drain off.

Transfer to a serving plate, cut into quarters, and serve.

Nutritional Analysis per Serving: calories 200, fat 8 g, protein 3 g, carbohydrates 29 g, sugar 16 g, fiber 3 g, sodium 150 mg

Sunchoke Gratin

Serves 4

In this gratin, the mellow flavor of the artichokes is nicely enriched by the yogurt and cheese. If you can't find Jerusalem artichokes, you could use artichoke hearts instead. Either way, this is a terrific supper dish served with a big salad.

2 tablespoons unsalted butter, preferably from grass-fed cows
1 tablespoon extra-virgin olive oil
2 large shallots, peeled and cut crosswise into thin slices
1 teaspoon minced garlic
450g Jerusalem artichokes, peeled and cut into thin slices
1 teaspoon fresh thyme leaves
1 teaspoon chopped fresh tarragon leaves
Sea salt and freshly ground black pepper
75–125ml vegetable stock or low-sodium vegetable broth
4 tablespoons organic, cultured, full-fat plain yogurt
50g mature cheddar cheese, grated

Preheat the grill.

Combine the butter and oil in a large frying pan over medium heat. When hot, add the shallots and garlic and cook, stirring occasionally, for about 6 minutes or just until soft and beginning to color.

Stir in the artichokes, thyme, and tarragon. Season with salt and pepper and add 75ml vegetable stock. Cover, reduce the heat to low, and simmer, stirring occasionally, for about 15 minutes or until the artichokes are very tender but not mushy. If the liquid evaporates, add a bit more stock. Uncover and continue to cook, stirring frequently, for about 4 minutes or until the artichokes are nicely glazed.

Remove the artichokes from the heat. Add the yogurt and gently stir to distribute evenly. Taste and, if necessary, season with additional salt and pepper. Sprinkle the cheese over the top and transfer to the grill.

Grill for about 3 minutes or until the cheese is melted and golden brown. Remove from the grill and serve immediately.

Nutritional Analysis per Serving: calories 222, fat 14 g, protein 6 g, carbohydrates 19 g, sugar 10 g, fiber 2 g, sodium 266 mg

Indian-Spiced Cabbage

Serves 6

A little spice added to sautéed cabbage lifts it from the ordinary to the sublime. If you don't like heat, feel free to eliminate the chilli. You may need to add a bit of water along with the cabbage to keep it from browning too quickly. But don't add too much, as a bit of color adds caramelization and savoriness to the mix.

3 tablespoons ghee or clarified butter, preferably from grass-fed cows

1 teaspoon mustard seeds

1 tablespoon minced garlic

1 teaspoon ground turmeric

¼ teaspoon ground cumin

675g red or green cabbage, trimmed, cored, and shredded

1 small green chilli, stemmed, seeded, and minced

Sea salt

Heat the ghee in a large frying pan over medium heat. Add the mustard seeds, cover, and cook for a couple of minutes, just until the seeds begin to pop.

Remove from the heat, uncover, and stir in the garlic, turmeric, and cumin. Return the pan to medium heat and cook, stirring constantly, for about 2 minutes to soften the garlic somewhat. Add the cabbage, chilli, and salt. Cook, tossing and turning, for a minute or so or until all of the cabbage is lightly coated with the seasoned ghee. Cover and cook for about 5 minutes or until the cabbage is still slightly crisp; if you prefer well-done cabbage, cook for an additional 20 minutes or until it is very soft and almost mushy.

Remove from the heat and serve.

Nutritional Analysis per Serving: calories 102, fat 7 g, protein 2 g, carbohydrates 9 g, sugar 4 g, fiber 3 g, sodium 31 mg

Thai Vegetable Curry

Serves 4

Although you can purchase both red and green curry pastes that give that identifiable Thai flavor to curries, I prefer to make my own. It keeps very well and is great to have on hand for last-minute curries. If you want to keep the curry completely vegetarian, you can eliminate the shrimp paste and fish sauce from the recipe. Or, if you prefer, replace the fish sauce and shrimp paste with about 100g shredded wakame or other seaweed to give a hint of the sea without the flavor of seafood.

1 tablespoon coconut oil
75g chopped onion
1 teaspoon minced garlic
1 teaspoon chopped fresh ginger
3 tablespoons Red Curry Paste (recipe follows)
500ml vegetable stock or low-sodium vegetable broth
1 (400ml) can unsweetened coconut milk
1 small aubergine, trimmed and cut into small cubes
1 small red bell pepper, stemmed, seeded, and cut into cubes
500g small broccoli florets
200g baby spinach, tough stems removed

Heat the oil in a large frying pan over medium heat. Add the onion, garlic, and ginger and cook, stirring frequently, for about 4 minutes or until soft. Add the curry paste, along with the stock and coconut milk, and bring to a simmer. Stir in the aubergine, bell pepper, and broccoli florets and cook, stirring frequently, for about 10 minutes or until the vegetables are just tender. Add the spinach and lower the heat. Cover and cook for 5 minutes or until the vegetables are very tender.

Serve immediately.

Nutritional Analysis per Serving: calories 290, fat 19 g, protein 7 g, carbohydrates 24 g, sugar 8 g, fiber 8 g, sodium 332 mg

Red Curry Paste

Makes about 250ml

Once you see how easy it is to make homemade red curry paste with this recipe, you'll never buy packaged curry pastes again. This Thai-inspired red curry paste is tastier, richer, and healthier than any you can buy in a store. It can be used in a variety of dishes, including those that feature seafood, poultry, and beef. You can also add a dollop of this paste to soups to add a punch of mouthwatering flavor.

10 dried red chillies, stemmed and seeded
250ml boiling water
10 black peppercorns
1 teaspoon caraway seeds, toasted
1 teaspoon coriander seeds, toasted
½ teaspoon ground turmeric
¼ teaspoon ground cinnamon
1 large shallot, finely chopped
2 tablespoons minced lemongrass or 1 tablespoon grated lemon zest
2 tablespoons fresh coriander leaves
1 tablespoon minced garlic
1 tablespoon shrimp paste
1 tablespoon additive-free fish sauce
1 teaspoon grated lime zest

Place the chillies in a heatproof container. Add the boiling water and set aside to rehydrate for 15 minutes. Drain well and pat dry.

Combine the drained chillies with the peppercorns, caraway seeds, coriander seeds, turmeric, and cinnamon in a spice grinder or the small jar of a blender. Process until finely ground.

Scrape the chilli mixture into the bowl of a food processor fitted with the metal blade. Add the shallot, lemongrass, coriander, garlic, shrimp paste, fish sauce, and lime zest, and process to a thick paste. If necessary, add cool water, a tablespoon or so at a time, to smooth out the mix.

Scrape the mixture from the processor bowl and place in a nonreactive container. Use immediately, or cover and refrigerate for up to 1 month.

Nutritional Analysis per Serving (1 tablespoon): calories 27, fat 0 g, protein 2 g, carbohydrates 4 g, sugar 0 g, fiber 0 g, sodium 210 mg

Broccoli, Mushroom, and Feta Toss

Serves 4

This one-pot meal is quick to put together and cook; nothing is easier to put on the table after a long day at work. The broccoli can be replaced with a head of cauliflower and the feta with almost any semi-soft or hard cheese you like.

1 head broccoli

2 tablespoons extra-virgin olive oil

1 tablespoon unsalted butter, preferably from grass-fed cows

350g mushrooms, cleaned, stems removed, and sliced

1 teaspoon minced garlic

Sea salt and freshly ground black pepper

225g feta cheese, crumbled

2 tablespoons chopped fresh basil

Cut the broccoli into florets. Trim the outer tough skin from the stalks and then cut the stalks, crosswise, into thin coins. Set aside.

Preheat the grill.

Heat the oil and butter in a large frying pan over medium heat. Add the mushrooms and garlic and cook, stirring occasionally, for about 10 minutes or until the mushrooms begin to exude their liquid and brown around the edges.

Add the broccoli florets and stems and continue to cook, stirring frequently, for another 5 minutes or until the broccoli is crisp-tender. Season with salt and pepper.

Add the feta and basil, stirring to blend. Cover and cook for about 2 minutes or just until the cheese has begun to melt.

Remove from the heat and place under the grill for a couple of minutes to brown slightly. Remove from the grill and serve immediately.

Nutritional Analysis per Serving: calories 300, fat 20 g, protein 15 g, carbohydrates 25 g, sugar 5 g, fiber 6 g, sodium 830 mg

Mixed Vegetable Kraut

Makes 2 liters

Time required to prepare: 1 week

The addition of kale and chilli to the traditional cabbage makes this mixture particularly high in vitamin C, with the liquid as nutritious as the vegetables. Even just a tablespoon will boost your daily intake of the vitamin. For a hint of sourness and even more nutritional value, you can add the juice and zest of 1 Meyer lemon, and for added sweetness, the juice and zest of 1 medium orange. Use the kraut as a side dish for grilled meats, fish, or poultry; toss it into mixed greens as a salad; or eat it as a snack.

450g shredded cabbage

450g jicama or white radish, peeled and shredded

100g shredded kale

75g shredded Granny Smith apple

115g shredded leek, white part only

1 teaspoon minced garlic

1 teaspoon minced hot red chilli

1½ teaspoons fine sea salt, preferably fine Himalayan pink salt

4 tablespoons whey, or 1 package vegetable starter culture

Distilled water, as needed

Combine the cabbage, jicama, kale, apple, leek, garlic, and chilli in a large bowl, tossing to blend. Add the salt and, using your hands, begin massaging the salt into the vegetable mixture, working until the vegetables exude some liquid.

Pack an equal amount of the vegetable mixture and the liquid it has exuded into each of two sterilized 1-liter glass canning jars with clean, unused lids or two 1-liter crockpots with tight-fitting lids. Using your fingertips, a smaller jar or glass that will fit down into the larger jar, or a potato masher, press the mixture down as firmly as you can to allow the liquid to rise up and cover the vegetables. Add 2 tablespoons of the whey to each jar, leaving 2.5–5cm of space between the vegetables and the top of the jar to give them room to expand as they ferment. If the liquid and the whey do not cover the vegetables entirely, add enough cool distilled water to completely cover.

Place a bit of cool water into a small resealable plastic bag. You need just enough water to create a weight to keep the vegetables under the liquid. Seal the bag, eliminating all the air inside, place it on top of the vegetables, and push it down to ensure that the water bag is serving as a weight. Place the lid on the container and seal tightly.

Set aside in a cool, dark spot for 1 week. Check the fermentation process daily to make sure that the vegetables have remained covered with liquid. If the liquid level is low, remove the water bag and set it aside. Remove and discard any scum or mold that has formed, noting that it is not harmful, just unappetizing. Add distilled water to cover. Push the vegetables back down into the liquid, place the water bag on top to press them down, seal tightly, and set aside as before.

After 1 week, the kraut will be ready to eat, but it may also be transferred to the refrigerator and stored for up to 9 months.

Nutritional Analysis per Serving (125ml): calories 30, fat 0 g, protein 2 g, carbohydrates 7 g, sugar 2 g, fiber 1 g, sodium 230 mg

Asian-Scented Greens

Makes 1 liter

Time required to prepare: 3 days

Pickled greens, most frequently mustard, are served throughout Asia—alone or as components of soups, stews, or rice dishes. This recipe carries quite a bit of flavor with the combination of peppery greens, hot chillies, and aromatic ginger and garlic. All the ingredients offer health benefits, and the fermentation makes the whole batch even better for you.

225g dandelion or mustard greens, or kale

1 tablespoon slivered fresh ginger

1 teaspoon slivered garlic

2 hot red or green chillies, cut in half lengthwise

500ml distilled water, plus more as needed

4 tablespoons natural apple cider vinegar

2 tablespoons coconut sugar (see Note)

1 tablespoon fine sea salt

3 star anise

Trim the leaves from the stems of the greens. Cut the stems crosswise into 5cm-long pieces and chop the leaves. Pack the cut stems into a 1 liter glass measuring jug and then add enough chopped leaves to fill the measuring jug when packed down gently. Transfer the greens mixture to a bowl. Add the ginger and garlic, tossing to blend well. Then pack the greens mixture into a clean, sterilized container, such as a 1-liter glass canning jar with a clean, unused lid or a 1-liter crock with a tight-fitting lid, randomly placing the chillies among the greens as you go.

Combine the distilled water, vinegar, sugar, and salt in a small saucepan over medium heat. Bring to a boil, then immediately remove from the heat.

Add the star anise and let the brine cool for 3 minutes. Pour the hot brine over the greens, taking care that the greens are completely covered. Leave 2.5–5cm of space between the greens and the top of the jar to give them room to expand as they ferment. If there is not enough liquid to cover it, add enough cool distilled water to completely cover.

Place a bit of cool water into a small resealable plastic bag. You need just enough water to create a weight to keep the greens under the liquid. Seal the bag, eliminating all the air inside, place it on top of the greens, and push it down to ensure that the water bag is serving as a weight. Place the lid on the container and seal tightly. Transfer to the refrigerator and allow to ferment for 3 days before serving. The greens may be stored, refrigerated, for up to 6 months.

NOTE: Coconut sugar is available at health food stores, at specialty markets, at some supermarkets, and online.

Nutritional Analysis per Serving (125ml): calories 25, fat 0 g, protein 0 g, carbohydrates 6 g, sugar 2 g, fiber 1 g, sodium 600 mg

FISH

Herb-Roasted Wild Salmon

Serves 4

About as simple as you can get, but also as elegant. The salmon is a wonderful main course for a dinner party, as it cooks quickly and looks so inviting that the cook will be the star of the evening. Always purchase your salmon from a reputable fishmonger, as farmed salmon are often labeled wild. Note that a recent investigation by the conservation group Oceana found that about 43 percent of the salmon labeled in stores as wild was, in fact, farmed, so buyer beware.

1 tablespoon coconut oil

1 tablespoon fresh lemon juice

4 tablespoons chopped fresh herbs, such as parsley, tarragon, chervil, and/or dill, plus more for garnish

1 large shallot, finely chopped

1 (675g) 2.5cm-thick salmon fillet, skin and all pin bones removed

Sea salt and freshly ground black pepper

Lemon wedges, for garnish

Preheat the oven to 230°C/gas 8.

Place the oil and lemon juice in a baking pan large enough to hold the salmon. Transfer the pan to the oven and heat for about 4 minutes or until the oil is very hot.

Working quickly, remove the hot pan from the oven and stir in the herbs and shallot. Season the salmon with salt and pepper and add it to the pan. Carefully turn the salmon a couple of times to coat it with the herbs and liquid, ending with the skinned side down. Roast, basting two or three times, for about 10 minutes or just until the fish is slightly underdone in the center.

Remove from the oven and carefully transfer to a serving platter, spooning the pan juices over the fish. Garnish with the extra herbs and lemon wedges.

Serve immediately.

Nutritional Analysis per Serving: calories 240, fat 10 g, protein 34 g, carbohydrates 2 g, sugar 1 g, fiber 0 g, sodium 230 mg

Steamed Wild Salmon with Sautéed Leeks and Chard

Serves 4

The sautéed chard makes a colorful base for the pink salmon, but you can use almost any green that is in season. In the spring, dandelion greens will give a slightly bitter contrast to the rich, succulent fish.

2 tablespoons unsalted butter, preferably from grass-fed cows, melted
4 (175g) skinless, boneless wild salmon fillets
Sea salt and freshly ground black pepper
8 thin slices lemon
2 tablespoons extra-virgin olive oil, plus more for drizzling
450g thinly sliced leeks, white part only
675g chopped rainbow chard, tough ends removed

Preheat the oven to 230°C/gas 8.

Place a wire rack large enough to hold the salmon in a rimmed baking sheet. Set aside.

Cut four 25cm square pieces of parchment paper. Using a pastry brush, lightly coat the paper with the melted butter. Set aside.

Lightly season the salmon with salt and pepper. Place a slice of lemon on each piece of paper, top with a piece of seasoned salmon, and top the salmon with another slice of lemon. Tightly wrap each piece of paper around the salmon by folding in the seam and twisting the ends together. Place the wrapped salmon pieces on the rack in the prepared baking sheet.

Place in the oven and allow to steam in the parchment paper for about 8 minutes or just until the fish is slightly underdone in the center.

While the salmon is steaming, heat the olive oil in a large sauté pan over medium-high heat. Add the leeks and sauté for about 4 minutes or until soft but not colored. Add the chard and, using tongs, cook, tossing and turning, for about 4 more minutes or until the leeks and chard are tender. Season with salt and pepper and remove from the heat. Tent lightly with parchment paper to keep warm.

Remove the salmon from the oven and carefully open the paper packets. Be cautious, as the steam will be very hot.

Spoon an equal portion of the chard mixture in the center of each of four plates. Place a piece of steamed salmon on top of each chard mound. Drizzle with olive oil and serve immediately.

Nutritional Analysis per Serving: calories 324, fat 19 g, protein 35 g, carbohydrates 3 g, sugar 0 g, fiber 0 g, sodium 330 mg

Fish Fillets with Black Olives, Artichokes, and Shaved Brussels Sprout Slaw

Serves 2

Chef Fabrizio Aielli, of Sea Salt restaurant, brings us this elegant and tasty dish that uses the local catch of the day. You could use sea bream in lieu of the snapper. Find what's fresh in your area. Feel free to double this recipe for a party of four.

2 (175g) snapper fillets

Sea salt and freshly ground black pepper

4 tablespoons extra-virgin olive oil

2 cloves garlic, smashed

2 sprigs fresh rosemary, chopped

Juice from half a lemon

2 artichokes preserved in oil, quartered

12 pitted kalamata olives

8 tablespoons Shaved Brussels Sprout Slaw (recipe follows)

Season the fish with salt and pepper. Heat a large frying pan over medium-high heat. Add the olive oil and bring to about its smoking point. Add the fish fillets to the pan, skin side down. Turn the heat down to medium and cook for 2 minutes. Using a fish spatula, flip the fillets and add the garlic, rosemary, and lemon juice. Cook for 2 more minutes or until the desired doneness is reached and the fish flakes easily with a fork. Remove the fish from the pan and place on two plates.

To the same pan, add the artichokes and olives and cook for 1 minute. Scatter around the fish, and top each with the slaw. Serve immediately.

Nutritional Analysis per Serving: calories 625, fat 44 g, protein 40 g, carbohydrates 23 g, sugar 3 g, fiber 12 g, sodium 670 mg

Shaved Brussels Sprout Slaw

Serves 2

This delicious slaw goes well with fish dishes. Double or triple the recipe if you're serving more people. You can also store the ingredients separately in airtight containers in the refrigerator and dress the slaw just prior to serving.

225g brussels sprouts

2 tablespoons liquid olive dressing (see Note)

Thinly shave the brussels sprouts on a mandoline. Toss with the liquid olive dressing. Serve.

Note: To make the liquid olive oil dressing, simply whisk together 1 large egg yolk and 125ml extra-virgin olive oil, adding the oil a little at a time until fully incorporated. Add a squeeze of fresh lemon juice and sea salt to taste. Double or triple the recipe to have the dressing on hand for several days. Store in an airtight container in the refrigerator for up to 1 week.

Nutritional Analysis per Serving: calories 290, fat 29 g, protein 4 g, carbohydrates 8 g, sugar 2 g, fiber 3 g, sodium 170 mg

MEAT AND POULTRY

Grass-Fed Beef Burgers

Serves 4

Rather than using just plain beef, I like to spice up my minced beef with some heat. Make sure that your beef is not too lean, as you need a good amount of fat to create a juicy, flavorful burger. For an extra treat, sauté some onions in butter until they are just beginning to soften; then, pile them on the grilled burger.

675g minced grass-fed beef

1 serrano or other hot green chilli, stemmed, seeded, and minced, or to taste

2 tablespoons minced shallot

1 teaspoon minced garlic

Sea salt and freshly ground black pepper

Extra-virgin olive oil, for brushing

Preheat and oil the grill or preheat a stovetop griddle pan over medium-high heat.

Combine the beef with the chilli, shallot, and garlic in a medium bowl. Using your hands, squish together to blend well. Season with salt and pepper.

Divide the mixture into quarters, then shape each portion into a patty of equal size so they will cook evenly. Using a pastry brush, generously coat the outside of the patties with the olive oil.

Place the burgers on the grill and grill for 4 minutes. Turn and grill for another 4 minutes for medium-rare.

Remove from the grill and serve immediately.

Nutritional Analysis per Serving: calories 350, fat 24 g, protein 33 g, carbohydrates 1 g, sugar 0 g, fiber 0 g, sodium 400 mg

Roast Leg of Grass-Fed Lamb

Serves 6

I think everyone has a favorite way of roasting a leg of lamb—mine is quite simple. I make any number of slits in the meat and then fill each one with a clove of garlic. Not only does it scent the meat as it roasts, but the garlic also adds flavor to the pan juices, further enriching the sauce.

4 tablespoons extra-virgin olive oil
Juice and zest of 1 lemon
1 tablespoon chopped fresh rosemary
2 teaspoons fresh thyme leaves
1 (2.7kg) leg of grass-fed lamb
About 20 cloves garlic, peeled and, if large, cut in half
Sea salt and freshly ground black pepper
3 leeks, finely chopped, white part only
125g chicken stock or low-sodium chicken broth
4 tablespoons dry white wine
50g unsalted butter, preferably from grass-fed cows, at room temperature

Preheat the oven to 230°C/gas 8. Place a rack in a roasting pan large enough to hold the lamb and set aside.

Combine the oil with the lemon juice and zest, rosemary, and thyme in a small bowl.

Using a small sharp knife, make 20 small slits in random spots all over the lamb. Fill each slit with a piece of garlic. Using your hands, generously coat the outside of the lamb with the oil mixture, patting it into the meat. Generously season with salt and pepper.

Place the seasoned meat on the rack in the roasting pan. Transfer to the oven and roast for 40 minutes. Reduce the oven temperature to 190°C/gas 5 and continue to roast for another hour or until an instant-read thermometer inserted into the thickest part reads 57°C for medium-rare (or 65°C for medium).

Transfer the lamb to a cutting board, tent with parchment paper, and allow to rest for 10 minutes before carving. Note that the lamb will continue to cook while it rests, increasing the internal temperature by about 10 degrees.

Transfer the roasting pan to the stove top over medium heat. Add the leeks and cook, stirring up the browned bits from the bottom of the pan, for about 3 minutes. Stir in the stock and wine and bring to a boil. Boil, stirring frequently, for about 3 minutes or until the liquid has reduced somewhat. Add the butter and cook, stirring, for about 3 minutes or until a rich sauce has formed. Taste and, if necessary, season with salt and pepper.

Using a carving knife, cut the lamb into thin slices and place on a serving platter. Drizzle some of the sauce over the top and serve with the remaining sauce on the side.

Nutritional Analysis per Serving: calories 540, fat 29 g, protein 58 g, carbohydrates 5 g, sugar 0 g, fiber 0 g, sodium 550 mg

Tuscan-Style Pork Roast

Serves 6

Like many Italian recipes, traditional or inspired, this is a very simple dish to prepare, but it requires superb ingredients. Pasture-raised pork is now usually a heritage breed that has been allowed to roam freely in pastures and woods. It is richer in flavor than farmed pork, but it can also be almost as lean. With the heritage breeds, I prefer the Berkshire for its high fat content and juiciness when cooked.

1 (1.1kg) boneless pasture-raised pork loin, preferably with a layer of fat

4 tablespoons extra-virgin olive oil

10 juniper berries, crushed

8 cloves garlic, minced

1 tablespoon dried rosemary

1 tablespoon cracked black pepper

Sea salt

250ml chicken stock or low-sodium chicken broth

225g thinly sliced onions

135g thinly sliced fennel

Zest of 1 orange

1 teaspoon chopped fresh rosemary

Preheat the oven to 200°C/gas 6.

Place a rack in a roasting pan large enough to hold the pork and set aside.

Place the pork on a cutting board. Combine the oil with the juniper berries, garlic, dried rosemary, and cracked pepper in a small bowl. When well combined, rub the mixture over the pork, pressing it down to adhere to the meat and fat. Season with salt and transfer to the rack in the roasting pan, fat side up.

Place in the oven and roast for 45 minutes. Add the stock, onions, fennel, and orange zest and continue to roast for an additional 40 minutes or until an instant-read thermometer inserted into the thickest part reads 65°C for medium-well done.

Transfer the pork to a cutting board, tent with parchment paper, and allow to rest for 10 minutes before carving.

Using a sharp knife, cut the roast crosswise into slices. Spoon the onion gravy onto a serving platter and lay the slices, slightly overlapping, over the onions. Sprinkle with the fresh rosemary and serve immediately.

Nutritional Analysis per Serving: calories 270, fat 10 g, protein 37 g, carbohydrates 6 g, sugar 1 g, fiber 1 g, sodium 390 mg

Roasted Chicken Thighs with Parsley Sauce

Serves 4

If you keep hard-boiled eggs on hand, as I do, this is a quick and easy supper for a busy work night. Chicken thighs are quick to cook, juicy, and flavorful. The sauce is a classic, but combined with the roasted chicken, it creates a totally new and exciting dish.

8 bone-in, skin-on chicken thighs (about 900g)

125ml plus 2 tablespoons extra-virgin olive oil

Sea salt and freshly ground black pepper

3 hard-boiled egg yolks

1½ tablespoons white wine vinegar

3 tablespoons chopped fresh flat-leaf parsley

2 teaspoons minced shallot

Preheat the oven to 200°C/gas 6.

Place the chicken thighs in a baking pan or on a rimmed baking sheet. Drizzle with 2 tablespoons of the olive oil and season with salt and pepper. Transfer to the oven and roast, turning occasionally, for about 25 minutes or until golden brown and just cooked through.

While the chicken is roasting, make the sauce.

Combine the egg yolks and vinegar in the bowl of a food processor fitted with the metal blade and process until smooth. With the motor running, slowly add the remaining 125ml olive oil, processing until well emulsified.

Scrape the egg mixture into a small bowl. Stir in the parsley and shallot, season with salt and pepper, and stir to combine.

Remove the chicken thighs from the oven and place on a serving platter. Spoon some of the sauce over the top and serve any remaining sauce on the side.

Nutritional Analysis per Serving: calories 600, fat 52 g, protein 35 g, carbohydrates 1 g, sugar 0 g, fiber 0 g, sodium 450 mg

DESSERTS

Coconut Pudding

Serves 4

Chia seeds not only add nutrients and fiber to this dessert, but they also thicken it without the addition of starches. Unfortunately, they also need time to hydrate, so the pudding needs to be made a few hours in advance of serving.

- 250ml almond milk
- 2 teaspoons stevia
- 250ml unsweetened coconut milk
- 4 tablespoons white chia seeds
- ¼ teaspoon ground nutmeg
- 2 tablespoons unsweetened coconut flakes, toasted

Combine the almond milk and stevia in a medium bowl, whisking vigorously to incorporate. Add the coconut milk, chia seeds, and nutmeg, whisking to just combine.

Cover with plastic wrap and transfer to the refrigerator. Chill for at least 4 hours, whisking every hour for the first 4 hours to ensure that the seeds are fully hydrated. The pudding may be chilled for up to 24 hours before serving.

When ready to serve, sprinkle the top with the toasted coconut flakes.

Nutritional Analysis per Serving: calories 170, fat 15 g, protein 3 g, carbohydrates 7 g, sugar 1 g, fiber 4 g, sodium 66 mg

Easy Chocolate Mousse

Serves 6

This is so quick to make and is the equal to more elaborate mousse recipes in terms of its airiness and flavor. Although it can be refrigerated for a couple of days, it gets firmer and firmer as it sits. It will still be delicious, but it will have quite a different texture.

200g extra-dark (72% cacao) chocolate, finely chopped
500ml chilled double cream, preferably from grass-fed cows
4 tablespoons organic, cultured, full-fat plain yogurt (optional)
Chocolate shavings, for garnish (optional)

Place the chocolate in a heatproof bowl.

Select a saucepan that will hold the bowl snugly inside, like a double boiler. Fill the saucepan about half full with water, taking care that the water will not touch the bottom of the bowl. Place over medium-high heat and fit the bowl into it. Bring the water to a simmer, frequently stirring the chocolate. By the time the water comes to almost a boil, the chocolate should be melted. Ideally, the chocolate should register no more than 48°C on a candy thermometer. (If the chocolate is too hot, it will instantly melt the whipped cream. At 48°C you should be able to do a heat test with your finger without any discomfort. It will be more than lukewarm, but not hot.) Remove the bowl from the saucepan and, using a wooden spoon, vigorously beat for about 30 seconds to aerate and even out the temperature.

While the chocolate is melting, whip the cream. Place the cold cream into a chilled bowl and, using a handheld electric mixer, beat for about 4 minutes or until soft peaks form.

Whisking constantly, slowly pour the cold whipped cream into the warm melted chocolate, beating until well blended. The mixture will be soft and almost airy.

You can either scrape the mousse into one large serving bowl or spoon an equal portion into each of six individual dessert cups or bowls. Refrigerate for at least 30 minutes before serving.

Serve as is, or garnish with the yogurt and some chocolate shavings.

Nutritional Analysis per Serving: calories 500, fat 46 g, protein 5 g, carbohydrates 20 g, sugar 11 g, fiber 4 g, sodium 35 mg

Ricotta with Berries and Toasted Almonds

Serves 4

Another easy dessert that is very satisfying. I usually make my own ricotta so that I am assured of its quality and flavor, but you can also find high-quality ricotta in the markets today. The cheese is rich, while the berries add a touch of sweetness and the almonds a nice finishing crunch.

- 240g full-fat ricotta cheese, preferably from a grass-fed animal (cow, goat, or sheep)
- 120g raspberries, strawberries, or blueberries
- 4 teaspoons flaked almonds or unsweetened coconut flakes, toasted

Spoon a quarter of the ricotta into each of four small dessert bowls. Sprinkle a quarter of the berries over the ricotta in each bowl. Spoon an equal portion of the almonds over the top. Serve immediately.

Nutritional Analysis per Serving: calories 135, fat 9 g, protein 8 g, carbohydrates 7 g, sugar 0 g, fiber 2 g, sodium 52 mg

Acknowledgments

I am truly blessed and in deep gratitude to be working with what has proven to be the absolute dream team in publishing. It is through the creative and artistic dedication to the craft of writing that Kristin Loberg so skillfully weaves the raw material I deliver into empowering text that changes the health destiny of so many. Bonnie Solow, my literary agent, maintains the vision for the entire team. Her highly skilled and compassionate guidance has allowed all of our goals to manifest. And Tracy Behar, our Editor at Little, Brown, with her gentle demeanor coupled with unparalleled literary acumen and experience, has made the process of creating this work a joyful event for all involved. And to her entire crew: Michael Pietsch, Reagan Arthur, Nicole Dewey, Craig Young, Genevieve Nierman, Lisa Erickson, Kaitlyn Boudah, Zea Moscone, Ben Allen, Julianna Lee, Valerie Cimino, Giraud Lorber, Olivia Aylmer, Katy Isaacs, and Dianne Schneider.

James Murphy—for your incredible ability to embrace the big picture and then bring about the manifestation of our shared goals.

Andrew Luer—for your day-to-day commitment to our short- and long-term goals, your ability to adapt to the ever-changing demands placed upon us, and your thoughtful council, which I highly respect.

Digital Natives—for so effectively navigating the constantly shifting landscape of social media and keeping our message in the forefront.

Judith Choate—for making magic happen in the kitchen and

creating such delicious recipes that live up to the Grain Brain Whole Life Plan principles.

Gigi Stewart—for contributing some of your own tasty kitchen concoctions that abide by my rules and make cooking fun.

Fabrizio Aielli and Seamus Mullen—for providing me the recipes to some of my favorite dishes from your respective restaurants, Sea Salt in Naples, Florida, and Tertulia in New York City.

Jonathan Heindemause—for performing the nutritional analyses on the recipes and making yourself available to handle all the last-minute changes to the menu.

Nicole Dunn—for hitting the ground running as the newest member to our team, and doing an amazing job in public relations.

And finally, to all of those who have inspired, helped, and supported me on my own journey. You know who you are. Thank you.

Selected Bibliography

The following is a selected list of papers and writings that have been useful in crafting this book, organized by chapter. This list is by no means exhaustive, since each of these entries could be complemented with dozens if not hundreds of others, but it will help you learn more and live up to the lessons and principles of *The Grain Brain Whole Life Plan*. This bibliography can also open other doors for further research and inquiry. For additional references and resources, please visit www.DrPelrmutter.com.

Introduction: You've Come to This Book for a Reason

Roach, Michael, and Christie McNally. *How Yoga Works*. New Jersey: Diamond Cutter Press, 2005.

Part I: Welcome to the Grain Brain Whole Life Plan

Chapter 1: What Is the Grain Brain Whole Life Plan?

Alzheimer's Association, "2016 Alzheimer's Disease Facts and Figures." www.alz.org/facts/ (accessed July 6, 2016).

Bournemouth University. "Brain Diseases Affecting More People and Starting Earlier Than Ever Before." ScienceDaily. www.sciencedaily.com/releases/2013/05/130510075502.htm (accessed June 14, 2016).

Centers for Disease Control and Prevention, Chronic Disease Prevention and Health Promotion. "Statistics and Tracking." www.cdc.gov/chronicdisease/stats/ (accessed June 14, 2016).

Centers for Disease Control and Prevention, National Center for Health Statistics. "Leading Causes of Death." www.cdc.gov/nchs/fastats/leading-causes-of-death.htm (accessed June 14, 2016).

Keith, Lierre. *The Vegetarian Myth: Food, Justice, and Sustainability*. Oakland, CA: PM Press, 2009.

Laidman, Jenni. "Obesity's Toll: 1 in 5 Deaths Linked to Excess Weight." Medscape.com. www.medscape.com/viewarticle/809516 (accessed June 10, 2016).

Perlmutter, David. "Bugs Are Your Brain's Best Friends." *Extraordinary Health*. Vol 24, 2015: 9–13.

Pritchard, C., A. Mayers, and D. Baldwin. "Changing Patterns of Neurological Mortality in the 10 Major Developed Countries 1979–2010." *Publ. Health* 127, no. 4 (2013): 357–368.

Chapter 2: The Chief Goals

Blumberg, R., and F. Powrie. "Microbiota, Disease, and Back to Health: A Metastable Journey." *Sci. Transl. Med.* 4, no. 137 (June 2012): 137rv7.

Braniste, V. et al. "The Gut Microbiota Influences Blood-Brain Barrier Permeability in Mice." *Sci. Transl. Med.* 6, no. 263 (November 2014): 263ra158.

Brogan, Kelly. *A Mind of Your Own: The Truth About Depression and How Women Can Heal Their Bodies to Reclaim Their Lives*. New York, NY: HarperWave, 2016.

Cahill Jr., G. F. and R. L. Veech. "Ketoacids? Good Medicine?" *Trans. Am. Clin. Climatol. Assoc.* 114 (2003): 149–61; discussion 162–3.

Carding, S. et al. "Dysbiosis of the Gut Microbiota in Disease." *Microb. Ecol. Health. Dis.* 26 (February 2015): 26191.

Cheema, A. K. et al. "Chemopreventive Metabolites Are Correlated with a Change in Intestinal Microbiota Measured in A-T Mice and Decreased Carcinogenesis." *PLoS One* 11, no. 4 (April 2016): e0151190.

Crane, P. K. et al. "Glucose Levels and Risk of Dementia." *N. Engl. J. Med.* 369, no. 6 (August 2013): 540–8.

Daulatzai, M. A. "Obesity and Gut's Dysbiosis Promote Neuroinflammation, Cognitive Impairment, and Vulnerability to Alzheimer's Disease: New Directions and Therapeutic Implications." *J. Mol. Gen. Med.* S1 (2014).

David, L. A. et al. "Diet Rapidly and Reproducibly Alters the Human Gut Microbiome." *Nature* 505, no. 7484 (January 2014): 559–63.

Earle, K. A. et al. "Quantitative Imaging of Gut Microbiota Spatial Organization." *Cell Host Microbe* 18, no. 4 (October 2015): 478–88.

Fan, Shelly. "The Fat-Fueled Brain: Unnatural or Advantageous?" ScientificAmerican.com (Mind Guest Blog). blogs.scientificamerican.com/mind-guest-blog/the-fat-fueled-brain-unnatural-or-advantageous/ (accessed June 10, 2016).

Gao, B. et al. "The Clinical Potential of Influencing Nrf2 Signaling in Degenerative and Immunological Disorders." *Clin. Pharmacol.* 6 (February 2014): 19–34.

Gedgaudas, Nora T. *Primal Body, Primal Mind: Beyond the Paleo Diet for Total Health and a Longer Life*. Rochester, Vermont: Health Arts Press, 2009.

Graf, D. et al. "Contribution of Diet to the Composition of the Human Gut Microbiota." *Microb. Ecol. Health. Dis.* 26 (February 2015): 26164.

Holmes, E. et al. "Therapeutic Modulation of Microbiota-Host Metabolic Interactions." *Sci. Transl. Med.* 4, no. 137 (June 2012): 137rv6.

Jones, R. M. et al. "Lactobacilli Modulate Epithelial Cytoprotection through the Nrf2 Pathway." *Cell Rep.* 12, no. 8 (August 2015): 1217–25.

Kelly, J. R. et al. "Breaking Down the Barriers: The Gut Microbiome, Intestinal Permeability and Stress-Related Psychiatric Disorders." *Front. Cell. Neurosci.* 9 (October 2015): 392.

Kresser, Chris. "9 Steps to Perfect Health—#5: Heal Your Gut." ChrisKresser.com. February 24, 2011. chriskresser.com/9-steps-to-perfect-health-5-heal-your-gut/ (accessed June 14, 2016).

Kumar, Himanshu et al. "Gut Microbiota as an Epigenetic Regulator: Pilot Study Based on Whole-Genome Methylation Analysis." *mBio* 5, no. 6 (December 2014): pii: e02113–14.

Li, H. et al. "The Outer Mucus Layer Hosts a Distinct Intestinal Microbial Niche." *Nat. Commun.* 6 (September 2015): 8292.

Mandal, Ananya. "History of the Ketogenic Diet." News-Medical.net. www.news-medical.net/health/History-of-the-Ketogenic-Diet .aspx (accessed June 14, 2016).

Mu, C. et al. "Gut Microbiota: The Brain Peacekeeper." *Front. Microbiol.* 7 (March 2016): 345.

Perlmutter, David. *Brain Maker: The Power of Gut Microbes to Heal and Protect Your Brain—For Life*. New York: Little, Brown and Co., 2015.

Perlmutter, David. "Why Eating for Your Microbiome is the Key to a Healthy Weight." MindBodyGreen.com guest blog. March 24, 2016. www.mindbodygreen.com/0-24285/why-eating-for-your-microbiome-is-the-key-to-a-healthy-weight.html (accessed June 14, 2016).

Reger, M. A. et al. "Effects of Beta-Hydroxybutyrate on Cognition in Memory-Impaired Adults." *Neurobiol. Aging* 25, no. 3 (March 2004): 311–4.

Rosenblat, J. D. et al. "Inflamed Moods: A Review of the Interactions Between Inflammation and Mood Disorders." *Prog. Neuropsychopharmacol. Biol. Psychiatry* 53 (August 2014): 23–34.

Schilling, M. A. "Unraveling Alzheimer's: Making Sense of the Relationship Between Diabetes and Alzheimer's Disease." *J. Alzheimers Dis.* 51, no. 4 (February 2016): 961–77.

Shenderov, B. A. "Gut Indigenous Microbiota and Epigenetics." *Microb. Ecol. Health Dis.* 23 (March 2012).

Slavin, Joanne. "Fiber and Prebiotics: Mechanisms and Health Benefits." *Nutrients* 5, no. 4 (April 2013): 1417–1435.

Sonnenburg, J. L., and M. A. Fischbach. "Community Health Care: Therapeutic Opportunities in the Human Microbiome." *Sci. Transl. Med.* 3, no. 78 (April 2011): 78ps12.

Stulberg, E. et al. "An Assessment of US Microbiome Research." *Nature Microbiology* 1, no. 15015 (January 2016).

Sunagawa, S. et al. "Ocean Plankton. Structure and Function of the Global Ocean Microbiome." *Science* 348, no. 6237 (May 2015): 1261359.

University of California—Los Angeles Health Sciences. "Gut Bacteria Could Help Prevent Cancer." ScienceDaily. www.sciencedaily.com/releases/2016/04/160413151108.htm (accessed June 14, 2016).

Vojdani, A. et al. "The Prevalence of Antibodies Against Wheat and Milk Proteins in Blood Donors and Their Contribution to Neuroimmune Reactivities." *Nutrients* 6, no. 1 (December 2013): 15–36.

Zhan, Y. et al. "Telomere Length Shortening and Alzheimer's Disease—A Mendelian Randomization Study." *JAMA Neurol.* 72, no. 10 (October 2015): 1202–3.

Zonis, S. et al. "Chronic Intestinal Inflammation Alters Hippocampal Neurogenesis." *J. Neuroinflamm.* 12 (April 2015): 65.

Chapter 3: The Food Rules

"GMO Foods: What You Need to Know." *Consumer Reports*. March 2015.

Bawa, A. S. and K. R. Anilakumar. "Genetically Modified Foods: Safety, Risks and Public Concerns—A Review." *J. Food Sci. Technol.* 50, no. 6 (December 2013): 1035–46.

Bazzano, L. A. et al. "Effects of Low-Carbohydrate and Low-Fat Diets: A Randomized Trial." *Ann. Intern. Med.* 161, no. 5 (September 2014): 309–18.

Catassi, C. et al. "A Prospective, Double-Blind, Placebo-Controlled Trial to Establish a Safe Gluten Threshold for Patients with Celiac Disease." *Am. J. Clin. Nutr.* 85, no. 1 (January 2007): 160–6.

Catassi, C. et al. "Non-Celiac Gluten Sensitivity: The New Frontier of Gluten-Related Disorders." *Nutrients* 5, no. 10 (September 2013): 3839–53.

Di Sabatino, A. et al. "Small Amounts of Gluten in Subjects with Suspected Nonceliac Gluten Sensitivity: A Randomized, Double-Blind, Placebo-Controlled, Cross-Over Trial." *Clin. Gastroenterol. Hepatol.* 13, no. 9 (September 2015): 1604–12.e3.

Fasano, A. "Zonulin and Its Regulation of Intestinal Barrier Function: The Biological Door to Inflammation, Autoimmunity, and Cancer." *Physiol. Rev.* 91, no. 1 (January 2011): 151–75.

Guyton, K. Z. et al. "Carcinogenicity of Tetrachlorvinphos, Parathion, Malathion, Diazinon, and Glyphosate." *Lancet Oncol.* 16, no. 5 (May 2015): 490–1.

Hollon, J. et al. "Effect of Gliadin on Permeability of Intestinal Biopsy Explants from Celiac Disease Patients and Patients with Non-Celiac Gluten Sensitivity." *Nutrients* 7, no. 3 (February 2015): 1565–76.

Lawrence, G. D. "Dietary Fats and Health: Dietary Recommendations in the Context of Scientific Evidence." *Adv. Nutr.* 4, no. 3 (May 2013): 294–302.

Levine, M. E. et al. "Low Protein Intake Is Associated with a Major Reduction in IGF-1, Cancer, and Overall Mortality in the 65 and Younger but Not Older Population." *Cell. Metab.* 19, no. 3 (March 2014): 407–17.

Mason, Rosemary. "Glyphosate Is Destructor of Human Health and Biodiversity." Available at www.gmoevidence.com/dr-mason-glyphosate-is-destructor-of-human-health-and-biodiversity/ (accessed June 14, 2106).

Nierenberg, Cari. "How Much Protein Do You Need?" WebMD.com feature, Guide to a Healthy Kitchen. www.webmd.com/diet/healthy-kitchen-11/how-much-protein?page=2 (accessed June 14, 2016).

Pan, A. et al. "Red Meat Consumption and Mortality: Results from 2 Prospective Cohort Studies." *Arch. Intern. Med.* 172, no. 7 (April 2012): 555–63.

Perlmutter, David. *Grain Brain: The Surprising Truth about Wheat, Carbs, and Sugar—Your Brain's Silent Killers.* New York: Little, Brown and Co., 2013.

Shai, I. et al. "Weight Loss with a Low-Carbohydrate, Mediterranean, or Low-Fat Diet." *NEJM.* 359, no. 3 (July 2008): 229–241.

Suez, J. et al. "Artificial Sweeteners Induce Glucose Intolerance by Altering the Gut Microbiota." *Nature* 514, no. 7521 (October 2014): 181–6.

Thongprakaisang, S. et al. "Glyphosate Induces Human Breast Cancer Cells Growth via Estrogen Receptors." *Food Chem. Toxicol.* 59 (September 2013): 129–36.

Toledo, E. et al. "Mediterranean Diet and Invasive Breast Cancer Risk among Women at High Cardiovascular Risk in the PREDIMED Trial: A Randomized Clinical Trial." *JAMA Intern. Med.* 175, no. 11 (November 2015): 1752–60.

Valls-Pedret, C. "Mediterranean Diet and Age-Related Cognitive Decline: A Randomized Clinical Trial." *JAMA Intern. Med.* 175, no. 7 (July 2015): 1094–103.

Want, Liqun et al. "Lipopolysaccharide-Induced Inflammation Is Associated with Receptor for Advanced Glycation End Products in Human Endothelial Cells." *FASEB J.* 28, no. 1 (April 2104).

Part II: The Grain Brain Whole Life Plan Essentials

Chapter 4: Getting Started: Assess Your Risk Factors, Know Your Numbers, and Prepare Your Mind

Brandhorst, S. et al. "A Periodic Diet That Mimics Fasting, Promotes Multi-System Regeneration, Enhanced Cognitive Performance, and Healthspan." *Cell Metab.* 22, no. 1 (July 2015): 86–99.

Leslie, Mitch. "Short-Term Fasting May Improve Health." ScienceMagazine.org. June 18, 2015. www.sciencemag.org/news/2015/06/short-term-fasting-may-improve-health (accessed June 14, 2016).

Perlmutter, Austin. "5 Ways to Thrive While You Wean Off Carbohydrates." DrPerlmutter.com. www.drperlmutter.com/five-ways-thrive-wean-carbohydrates/ (accessed June 15, 2016).

Seshadri, S. et al. "Plasma Homocysteine As a Risk Factor for Dementia and Alzheimer's Disease." *N. Engl. J. Med.* 346, no. 7 (February 2002): 476–83.

Torgan, Carol. "Health Effects of a Diet That Mimics Fasting." National Institutes of Health, NIH Research Matters page on nih.gov. July 13, 2015. www.nih.gov/news-events/nih-research-matters/health-effects-diet-mimics-fasting (accessed June 15, 2016).

Youm, Y. H. et al. "The Ketone Metabolite β-hydroxybutyrate Blocks Nlrp3 Inflammasome-Mediated Inflammatory Disease." *Nat. Med.* 21, no. 3 (March 2015): 263–9.

Chapter 5: Step 1—Edit Your Diet and Pill-Popping

Azad, M. B. et al. "Infant Antibiotic Exposure and the Development of Childhood Overweight and Central Adiposity." *Int. J. Obes.* (Lond.) 38, no. 10 (October 2014): 1290–8.

Babiker, R. et al. "Effects of Gum Arabic Ingestion on Body Mass Index and Body Fat Percentage in Healthy Adult Females:

Two-Arm Randomized, Placebo Controlled, Double-Blind Trial." *Nutr. J.* 11 (December 2012): 111.

Björkhem, I., and S. Meaney. "Brain Cholesterol: Long Secret Life Behind a Barrier." *Arterioscler. Thromb. Vasc. Biol.* 24, no. 5 (May, 2004): 806–15.

Calame, W. et al. "Gum Arabic Establishes Prebiotic Functionality in Healthy Human Volunteers in a Dose-Dependent Manner." *Br. J. Nutr.* 100, no. 6 (December 2008): 1269–75.

Chowdhury, R. et al. "Vitamin D and Risk of Cause Specific Death: Systematic Review and Meta-Analysis of Observational Cohort and Randomised Intervention Studies." *BMJ.* 348 (April 2014): g1903.

Culver, A. L. et al. "Statin Use and Risk of Diabetes Mellitus in Postmenopausal Women in the Women's Health Initiative." *Arch. Intern. Med.* 172, no. 2 (January 23, 2012): 144–52.

Durso, G. R. et al. "Over-the-Counter Relief from Pains and Pleasures Alike: Acetaminophen Blunts Evaluation Sensitivity to Both Negative and Positive Stimuli." *Psychol. Sci.* 26, no. 6 (June 2015): 750–8.

Frenk, S. M. et al. "Prescription Opioid Analgesic Use Among Adults: United States, 1999–2012." NCHS Data Brief no. 189 (February 2015): 1–8.

Graham, D. Y. et al. "Visible Small-Intestinal Mucosal Injury in Chronic NDSAID Users." *Clin. Gastroenterol. Hepatol.* 3, no. 1 (January, 2005): 55–9.

Hegazy, G. A. et al. "The Role of Acacia Arabica Extract As an Antidiabetic, Antihyperlipidemic, and Antioxidant in Streptozotocin-Induced Diabetic Rats." *Saudi Med. J.* 34, no. 7 (July 2013): 727–33.

Holscher, H.D. et al. "Fiber Supplementation Influences Phylogenetic Structure and Functional Capacity of the Human Intestinal Microbiome: Follow-Up of a Randomized Controlled Trial." *Am. J. Clin. Nutr.* 101, no. 1 (January 2015): 55–64.

Kantor, E. D. et al. "Trends in Prescription Drug Use among Adults in the United States from 1999–2012." JAMA. 314, no. 17 (November 2015): 1818–31.

Kennedy, Pagan. "The Fat Drug." *New York Times*, Sunday Review. March 9, 2014, page SR1.

Lam, J. R. et al. "Proton Pump Inhibitor and Histamine 2 Receptor Antagonist Use and Vitamin B12 Deficiency." JAMA. 310, no. 22 (December 11, 2013): 2435–42.

Liew, A. et al. "Acetaminophen Use during Pregnancy, Behavioral Problems, and Hyperkinetic Disorders. JAMA *Pediatr.* 168, no. 4 (April, 2014): 313–20.

Littlejohns, T. J. et al. "Vitamin D and the Risk of Dementia and Alzheimer's Disease." *Neurology* 83, no. 10 (September 2014): 920–8.

Matthews, L. R. et al. "Worsening Severity of Vitamin D Deficiency Is Associated with Increased Length of Stay, Surgical Intensive Care Unit Cost, and Mortality Rate in Surgical Intensive Care Unit Patients." *Am. J. Surg.* 204, no. 1 (July 2012): 37–43.

Mazer-Amirshahi, M. et al. "Rising Rates of Proton Pump Inhibitor Prescribing in US Emergency Departments." *Am. J. Emerg. Med.* 32, no. 6 (June 2014): 618–22.

Mikkelsen, K. H. et al. "Use of Antibiotics and Risk of Type 2 Diabetes: A Population-Based Case-Control Study." *J. Clin. Endocrinol. Metab.* 100, no. 10 (October 2015): 3633–40.

Million, M. et al. "Correlation between Body Mass Index and Gut Concentrations of *Lactobacillus reuteri*, *Bifidobacterium animalis*, *Methanobrevibacter smithii* and *Escherichia coli*." *Int. J. Obes.* (Lond.) 37, no. 11 (November 2013): 1460–6.

Mor, A. et al. "Prenatal Exposure to Systemic Antibacterials and Overweight and Obesity in Danish Schoolchildren: A Prevalence Study." *Int. J. Obes.* (Lond.) 39, no. 10 (October 2015): 1450–5.

Newport, Mary. "What if There Was a Cure for Alzheimer's Disease and No One Knew?" CoconutKetones.com. July 22, 2008. www.coconutketones.com/whatifcure.pdf (accessed June 14, 2016).

Park, Alice. "Too Many Antibiotics May Make Children Heavier." Time.com. October 21, 2015. time.com/4082242/antibiotics-obesity/ (accessed June 14, 2016).

Pärtty, A. et al. "A Possible Link between Early Probiotic Intervention and the Risk of Neuropsychiatric Disorders Later in Childhood: A Randomized Trial." *Pediatr. Res.* 77, no. 6 (June 2015): 823–8.

Perlmutter, David. *Grain Brain: The Surprising Truth about Wheat, Carbs, and Sugar—Your Brain's Silent Killers.* New York: Little, Brown and Co., 2013.

Reyes-Izquierdo, T. et al. "Modulatory Effect of Coffee Fruit Extract on Plasma Levels of Brain-Derived Neurotrophic Factor in Healthy Subjects." *Br. J. Nutr.* 110, no. 3 (August 2013): 420–5.

Reyes-Izquierdo, T. et al. "Stimulatory Effect of Whole Coffee Fruit Concentrate Powder on Plasma Levels of Total and Exosomal Brain-Derived Neurotrophic Factor in Healthy Subjects: An Acute Within-Subject Clinical Study." *Food Nut. Sci.* 4, no. 9 (2013): 984–990.

Sass, Cynthia. "The 5 Most Confusing Health Labels." HuffingtonPost.com. www.huffingtonpost.com/2014/08/02/health-food-labels-confusing_n_5634184.html (accessed May 1, 2016).

Schwartz, B. S. et al. "Antibiotic Use and Childhood Body Mass Index Trajectory." *Int. J. Obes.* (Lond.) 40, no. 4 (April 2016): 615–21.

Shah, N. H. et al. "Proton Pump Inhibitor Usage and the Risk of Myocardial Infarction in the General Population." *PLoS One* 10, no. 6 (June 2015): e0124653.

Sigthorsson, G. et al. "Intestinal Permeability and Inflammation in Patients on NSAIDs." *Gut.* 43, no. 4 (October, 1998): 506–11.

Simakachorn, N. et al. "Tolerance, Safety, and Effect on the Faecal Microbiota of an Enteral Formula Supplemented with Pre- and Probiotics in Critically Ill Children." *J. Pediatr. Gastroenterol. Nutr.* 53, no. 2 (August 2011): 174–81.

Slavin, Joanne. "Fiber and Prebiotics: Mechanisms and Health Benefits." *Nutrients* 5, no. 4 (April 2013): 1417–1435.

Swaminathan, A., and G. A. Jicha. "Nutrition and Prevention of Alzheimer's Dementia." *Front. Aging. Neurosci.* 6 (October 2014): 282.

University of Exeter. "Link between Vitamin D, Dementia Risk Confirmed." ScienceDaily. www.sciencedaily.com/releases/2014/08/14 0806161659.htm (accessed June 15, 2016).

Velicer, C. M. et al. "Antibiotic Use in Relation to the Risk of Breast Cancer." *JAMA.* 291, no. 7 (February 2004): 827–35.

Vesper, B. J. et al. "The Effect of Proton Pump Inhibitors on the Human Microbiota." *Curr. Drug. Metab.* 10, no. 1 (January 2009): 84–9.

Weinstein, G. et al. "Serum Brain-Derived Neurotrophic Factor and the Risk for Dementia: The Framingham Heart Study." *JAMA Neurol.* 71, no. 1 (January 2014): 55–61.

World Health Organization. "WHO's First Global Report on Antibiotic Resistance Reveals Serious, Worldwide Threat to Public Health." WHO.int news release, April 30, 2014. www.who.int/mediacentre/news/releases/2014/amr-report/en/ (accessed June 14, 2016).

Wu, A. et al. "Curcumin Boosts DHA in the Brain: Implications for the Prevention of Anxiety Disorders." *Biochim. Biophys. Acta.* 1852, no. 5 (May 2015): 951–61.

Zaura, E. et al. "Same Exposure but Two Radically Different Responses to Antibiotics: Resilience of the Salivary Microbiome versus Long-Term Microbial Shifts in Feces." *mBio.* 6, no. 6 (November 2015): e01693–15.

Zhang, H. et al. "Discontinuation of Statins in Routine Care Settings: A Cohort Study." *Ann. Intern. Med.* 158, no. 7 (April 2, 2013): 526–34.

Chapter 6: Step 2—Add Your Support Strategies

American Academy of Neurology (AAN). "Heavy Snoring, Sleep Apnea May Signal Earlier Memory and Thinking Decline." ScienceDaily. www.sciencedaily.com/releases/2015/04/150415203338.htm (accessed June 15, 2016).

Andrews, S. et al. "Beyond Self-Report: Tools to Compare Estimated and Real-World Smartphone Use." *PLoS One* 10, no. 10 (October 2015): e0139004.

Balogun, J. A. et al. "Comparison of the EMG Activities in the Vastus Medialis Oblique and Vastus Lateralis Muscles During Hip Adduction and Terminal Knee Extension Exercise Protocols." *Afr. J. Physiother. and Rehab. Sci.* 2, no. 1 (2010).

Barclay, Eliza. "Eating to Break 100: Longevity Diet Tips from the Blue Zones." NPR.com. The Salt page. April 11, 2015. www.npr.org/sections/thesalt/2015/04/11/398325030/eating-to-break-100-longevity-diet-tips-from-the-blue-zones (accessed June 14, 2016).

Berman, M. G. et al. "Interacting with Nature Improves Cognition and Affect for Individuals with Depression." *J. Affect. Disord.* 140, no. 3 (November 2012): 300–5.

Buettner, Dan. "The Island Where People Forget to Die." *New York Times*, Sunday Magazine. October 28, 2012, page MM36.

Clarke, S. F. et al. "Exercise and Associated Dietary Extremes Impact on Gut Microbial Diversity." *Gut* 63, no. 12 (December 2014): 1913–20.

Dennis, Brady. "Nearly 60 Percent of Americans—The Highest Ever—Are Taking Prescription Drugs." *Washington Post*, To Your Health section. November 3, 2015. www.washingtonpost.com/news/to-your-health/wp/2015/11/03/more-americans-than-ever-are-taking-prescription-drugs/ (accessed June 14, 2016).

Dimeo, F. et al. "Benefits from Aerobic Exercise in Patients with Major

Depression: A Pilot Study." *Br. J. Sports Med.* 35, no. 2 (April 2001): 114–7.

Environmental Working Group. www.ewg.org. Research section and Consumer Guides.

Erickson, K. I. et al. "Exercise Training Increases Size of Hippocampus and Improves Memory." *Proc. Natl. Acad. Sci.* 108, no. 7 (February 2011): 3017–22.

Eriksson, P. S. et al. "Neurogenesis in the Adult Human Hippocampus." *Nat. Med.* 4, no. 11 (November 1998): 1313–7.

Halden, Rolf. "Epistemology of Contaminants of Emerging Concern and Literature Meta-Analysis." *J. Haz. Mat.* 282, no. 23 (January 2015): 2–9.

Jarrett, Christian. "How Expressing Gratitude Might Change Your Brain." NYMag.com. Science of Us section. January 7, 2016. nymag.com/scienceofus/2016/01/how-expressing-gratitude-change-your-brain.html (accessed June 14, 2016).

Kini, P. et al. "The Effects of Gratitude Expression on Neural Activity." *Neuroimage.* 128 (March 2016): 1–10.

Lautenschlager, N. T. et al. "Effect of Physical Activity on Cognitive Function in Older Adults at Risk for Alzheimer's Disease: A Randomized Trial." *JAMA.* 300, no. 9 (September 2008): 1027–37.

Lee, B. H., and Y. K. Kim. "The Roles of BDNF in the Pathophysiology of Major Depression and in Antidepressant Treatment." *Psychiatry Investig.* 7, no. 4 (December 2010): 231–5.

McCann, I. L., and D. S. Holmes. "Influence of Aerobic Exercise on Depression." *J. Pers. Soc. Psychol.* 46, no. 5 (May 1984): 1142–7.

National Sleep Foundation. www.sleepfoundation.org. Sleep Disorders and Sleep Topics.

Osorio, R. S. et al. "Sleep-Disordered Breathing Advances Cognitive Decline in the Elderly." *Neurology* 84, no. 19 (May 2015): 1964–71.

Perlmutter, David, and Alberto Villoldo. *Power Up Your Brain: The Neuroscience of Enlightenment*. New York: Hay House, 2011.

Preidt, Robert. "Bonding with Others May Be Crucial for Long-Term Health." U.S. News & World Report Health, January 8, 2016. health.usnews.com/health-news/articles/2016-01-08/bonding-with-others-may-be-crucial-for-long-term-health (accessed June 14, 2016).

Raji, C. A. et al. "Longitudinal Relationships between Caloric Expenditure and Gray Matter in the Cardiovascular Health Study." *J. Alzheimers Dis.* 52, no. 2 (March 2016): 719–29.

Richtel, Matt. "Digital Devices Deprive Brain of Needed Downtime." NYTimes.com. August 24, 2010. www.nytimes.com/2010/08/25/technology/25brain.html (accessed June 14, 2016).

Sandler, David. "Dumbbell Wide Row for Serious Back Muscle." Muscle & Fitness. www.muscleandfitness.com/workouts/backexercizes/dumbell-wide-row-serious-back-muscle (accessed July 19, 2016).

Srikanthan, P., and A. S. Karlamangla. "Muscle Mass Index as a Predictor of Longevity in Older Adults." *Am. J. Med.* 127, no. 6 (June 2014): 547–53.

University of California—Los Angeles Health Sciences. "Older Adults: Build Muscle and You'll Live Longer." ScienceDaily. www.sciencedaily.com/releases/2014/03/140314095102.htm (accessed June 15, 2016).

Weinstein, G. et al. "Serum Brain-Derived Neurotrophic Factor and the Risk for Dementia: The Framingham Heart Study." *JAMA Neurol.* 71, no. 1 (January 2014): 55–61.

Yang, Y. C. et al. "Social Relationships and Physiological Determinants of Longevity across the Human Life Span." *Proc. Natl. Acad. Sci.* 113, no. 3 (January 2016): 578–83.

Chapter 7: Step 3—Plan Accordingly

Garaulet, M. et al. "Timing of Food Intake Predicts Weight Loss Effectiveness." *Int. J. Obes.* (Lond.) 37, no. 4 (April 2013): 604–11.

Chapter 8: Troubleshooting

Azad, M. B. et al. "Gut Microbiota of Healthy Canadian Infants: Profiles by Mode of Delivery and Infant Diet at 4 Months." CMAJ. 185, no. 5 (March 2013): 385–94.

Blustein, J., and Jianmeng Liu. "Time to Consider the Risks of Caesarean Delivery for Long-Term Child Health." *BMJ.* 350 (June 2015): h2410.

Couzin-Frankel, Jennifer. "How to Give a C-Section Baby the Potential Benefits of Vaginal Birth." ScienceMag.org. February 1, 2016. www.sciencemag.org/news/2016/02/how-give-c-section-baby-potential-benefits-vaginal-birth (accessed June 14, 2016).

Mueller, N. T. et al. "The Infant Microbiome Development: Mom Matters." *Trends Mol. Med.* 2014. dx.doi.org/10.1016/j.molmed.2014.12.002.

Part III: Let's Eat!

Chapter 9: Final Reminders and Snack Ideas

Otto, M. C. et al. "Everything in Moderation—Dietary Diversity and Quality, Central Obesity and Risk of Diabetes." *PLoS One* 10, no. 10 (October 2015): e0141341.

University of Texas Health Science Center at Houston. "'Everything in Moderation' Diet Advice May Lead to Poor Metabolic Health in US Adults." ScienceDaily. www.sciencedaily.com/releases/2015/10/151030161347.htm (accessed June 15, 2016).

Index

About the Author

Dr David Perlmutter has dedicated his professional career to advancing the science underlying brain disorders, with an emphasis on disease prevention. He is a board-certified neurologist and Fellow of the American College of Nutrition, and has contributed extensively to the world medical literature with publications appearing in *The Journal of Neurosurgery*, *The Southern Medical Journal*, *Journal of Applied Nutrition*, and *Archives of Neurology*. Dr Perlmutter lectures worldwide and is a frequent speaker at symposia sponsored by such institutions as Harvard University, Columbia University, and New York University. He is a popular guest on national media programs and has appeared on *20/20*, *Larry King Live*, *CNN*, *Fox News*, *Fox and Friends*, *The Today Show*, *Oprah*, *Dr. Oz*, and *The CBS Early Show* and serves as a medical advisor to the *Dr. Oz* show. In 2002 Dr Perlmutter was the recipient of the Linus Pauling Award for his innovative approaches to neurological disorders and in addition was presented with the Denham Harmon Award for his pioneering work in the application of free radical science to clinical medicine. In 2006 he was named Clinician of the Year by the National Nutritional Foods Association, which was followed by the Humanitarian of the Year award from the American College of Nutrition in 2010 as well as the 2015 Media Award from the same institution. He is the author of seven books including the #1 *New York Times* bestseller *Grain Brain*, now published in twenty-seven languages.